SCAN THE CODE TO ACCESS YOUR FREE DIGITAL COPY OF THE NEUROANATOMY COLORING BOOK

SCAN ME

The Neuroanatomy Coloring Book features:

•**The most effective way to skyrocket your neuroanatomical knowledge, all while having fun!**

• Full coverage of the major systems of the human brain to provide context and reinforce visual recognition

• **25+ unique, easy-to-color pages of different neuroanatomical sections with their terminology**

• Large 8.5 by 11-inch single side paper so you can easily remove your coloring

•**Self-quizzing for each page, with convenient same-page answer keys**

THIS BOOK BELONGS TO

TO

_____

TABLE OF CONTENTS

YOGA POSES FOR BEGINNERS

TABLE OF CONTENTS

YOGA POSES FOR INTERMEDIATES

TABLE OF CONTENTS

YOGA POSES FOR EXPERTS

YOGA POSES FOR BEGINNERS

1. MOUNTAIN POSE

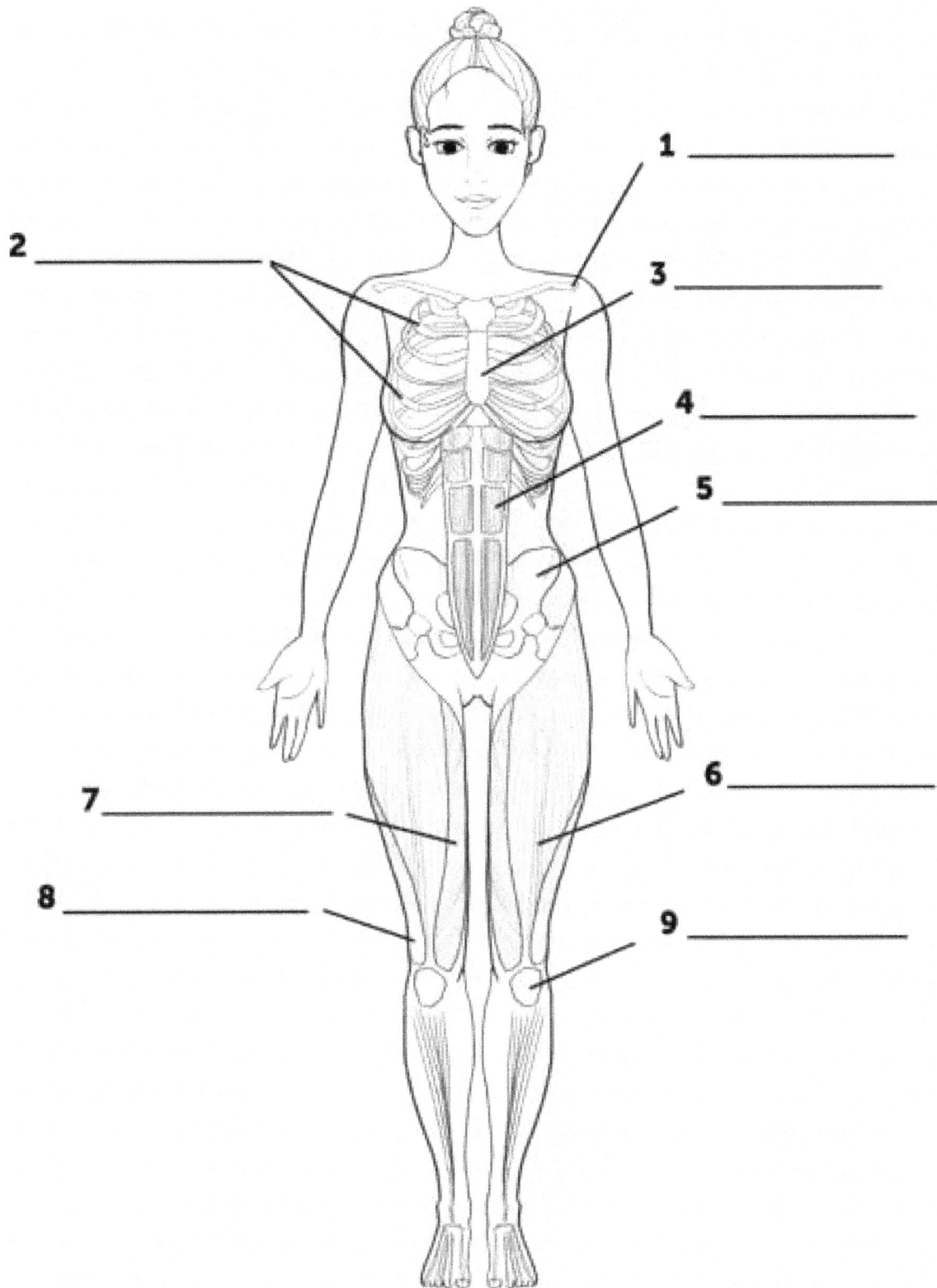

1 _____

2 _____

3 _____

4 _____

5 _____

6 _____

7 _____

8 _____

9 _____

1. MOUNTAIN POSE

1. COLLARBONE
2. RIBS
3. STERNUM
4. RECTUS ABDOMINIS
5. PELVIS
6. QUADRICEPS
7. VASTUS MEDIALIS
8. VASTUS LATERALIS
9. PATELLA

2. PALM TREE POSE

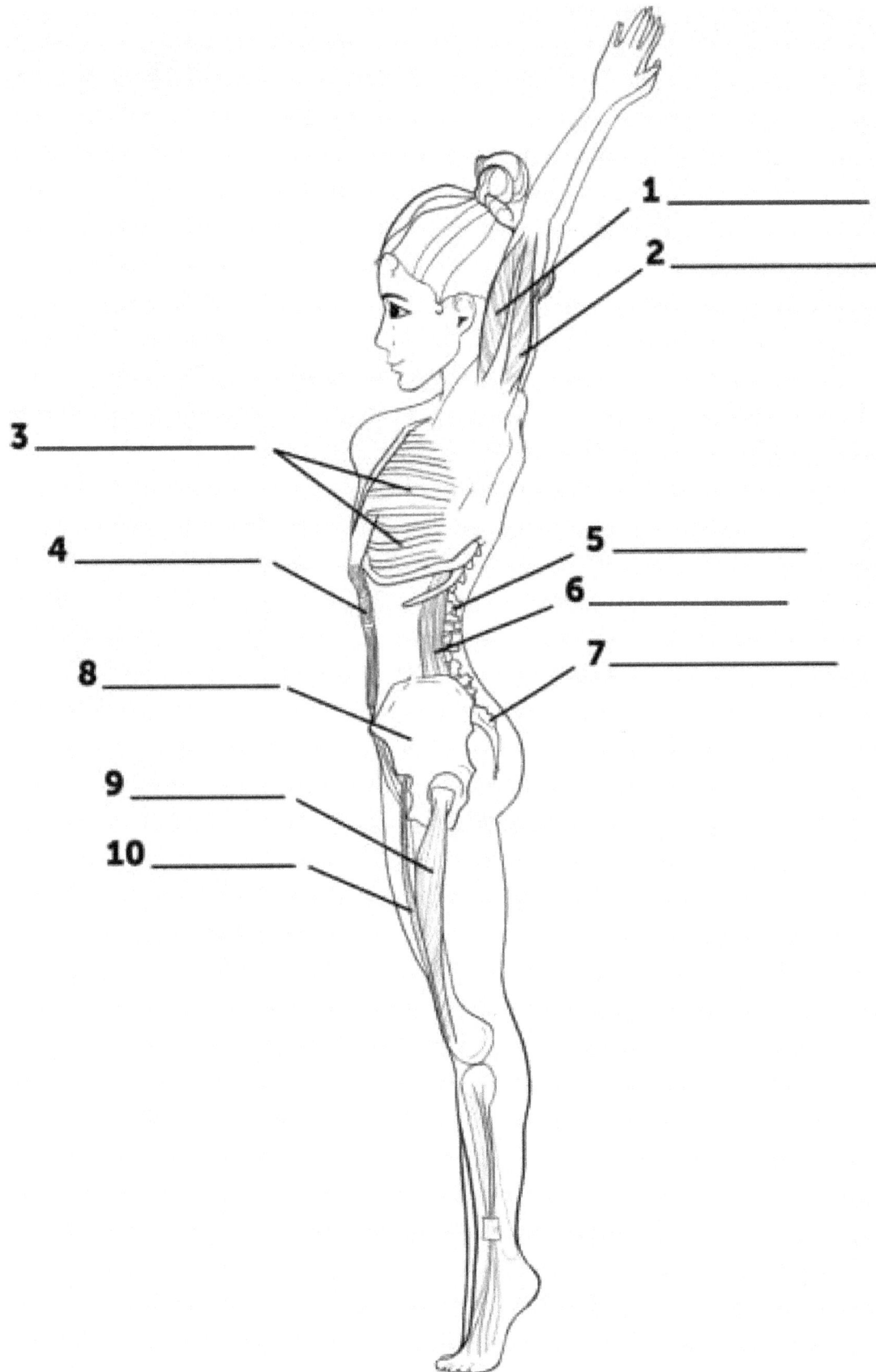

1 _____

2 _____

3 _____

4 _____

5 _____

6 _____

7 _____

8 _____

9 _____

10 _____

2. PALM TREE POSE

1. TRICEPS BRACHII
2. DELTOID
3. RIBS
4. RECTUS ABDOMINIS
5. SPINE
6. ERECTOR SPINAE
7. SACRUM
8. PELVIS
9. RECTUS FEMORIS
10. SARTORIUS

3. STANDING FORWARD BEND

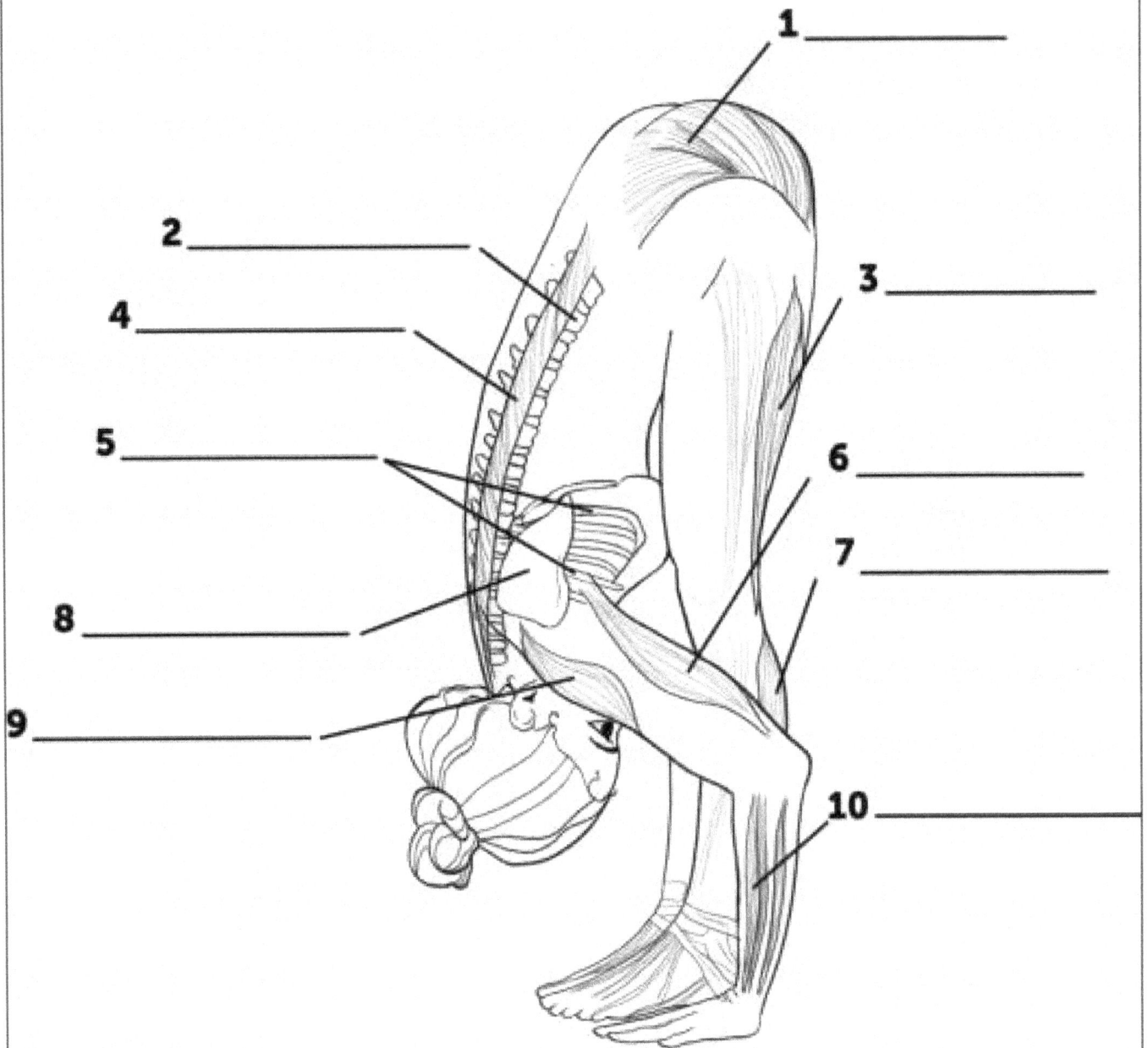

1 _____

2 _____

3 _____

4 _____

5 _____

6 _____

7 _____

8 _____

9 _____

10 _____

3. STANDING FORWARD BEND

1. PIRIFORMIS

2. SPINE

3. HAMSTRINGS

4. SPINAL MUSCLES

5. RIBS

6. TRICEPS BRACHII

7. GASTROCNEMIUS

8. SCAPULA

9. DELTOID

10. EXTENSOR DIGITORUM

4. HALF STANDING FORWARD BEND

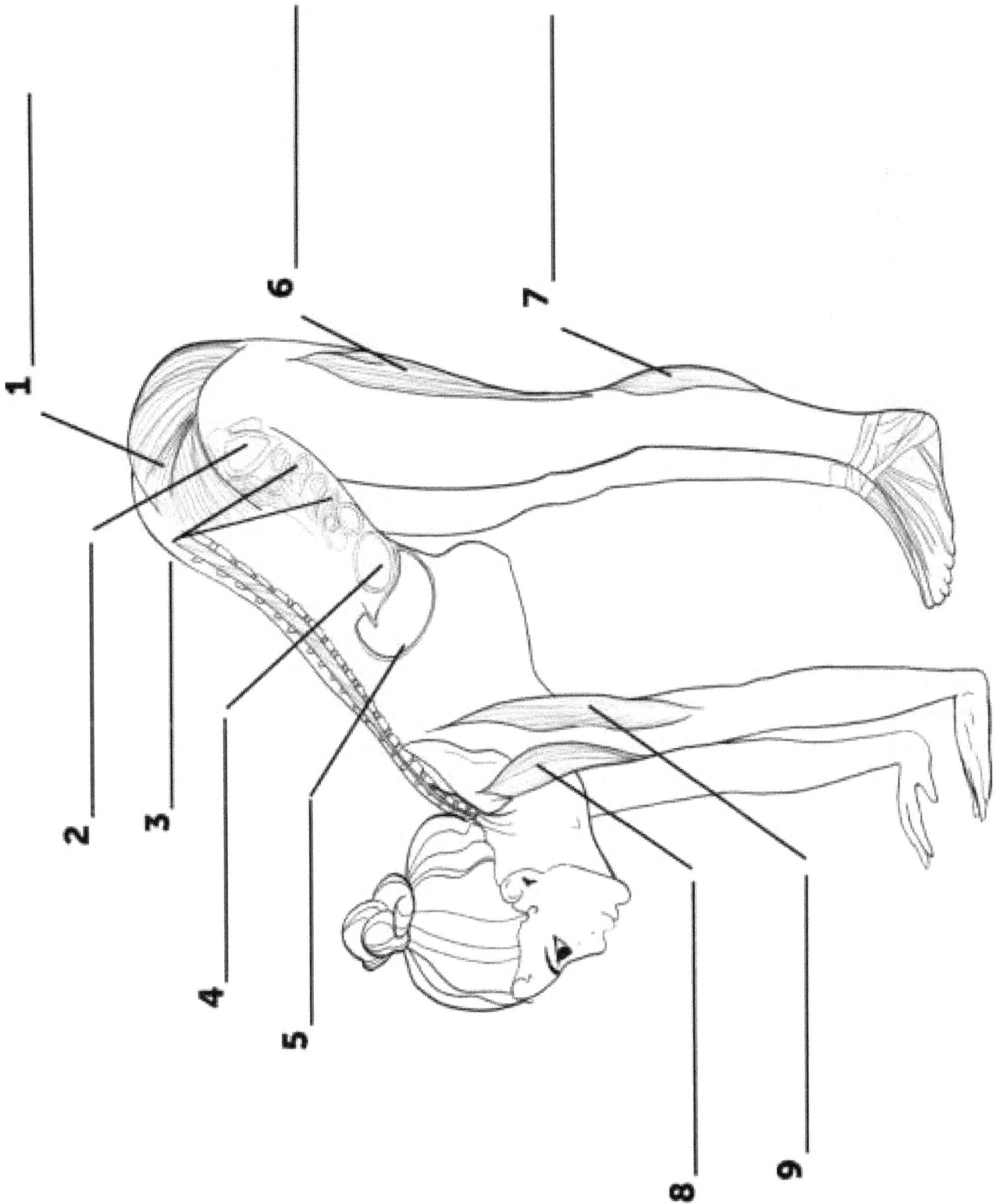

4. HALF STANDING FORWARD BEND

1. PIRIFORMIS
2. URINARY BLADDER
3. SMALL INTESTINE
4. STOMACH
5. LIVER
6. HAMSTRINGS
7. GASTROCNEMIUS
8. DELTOID
9. TRICEPS BRACHII

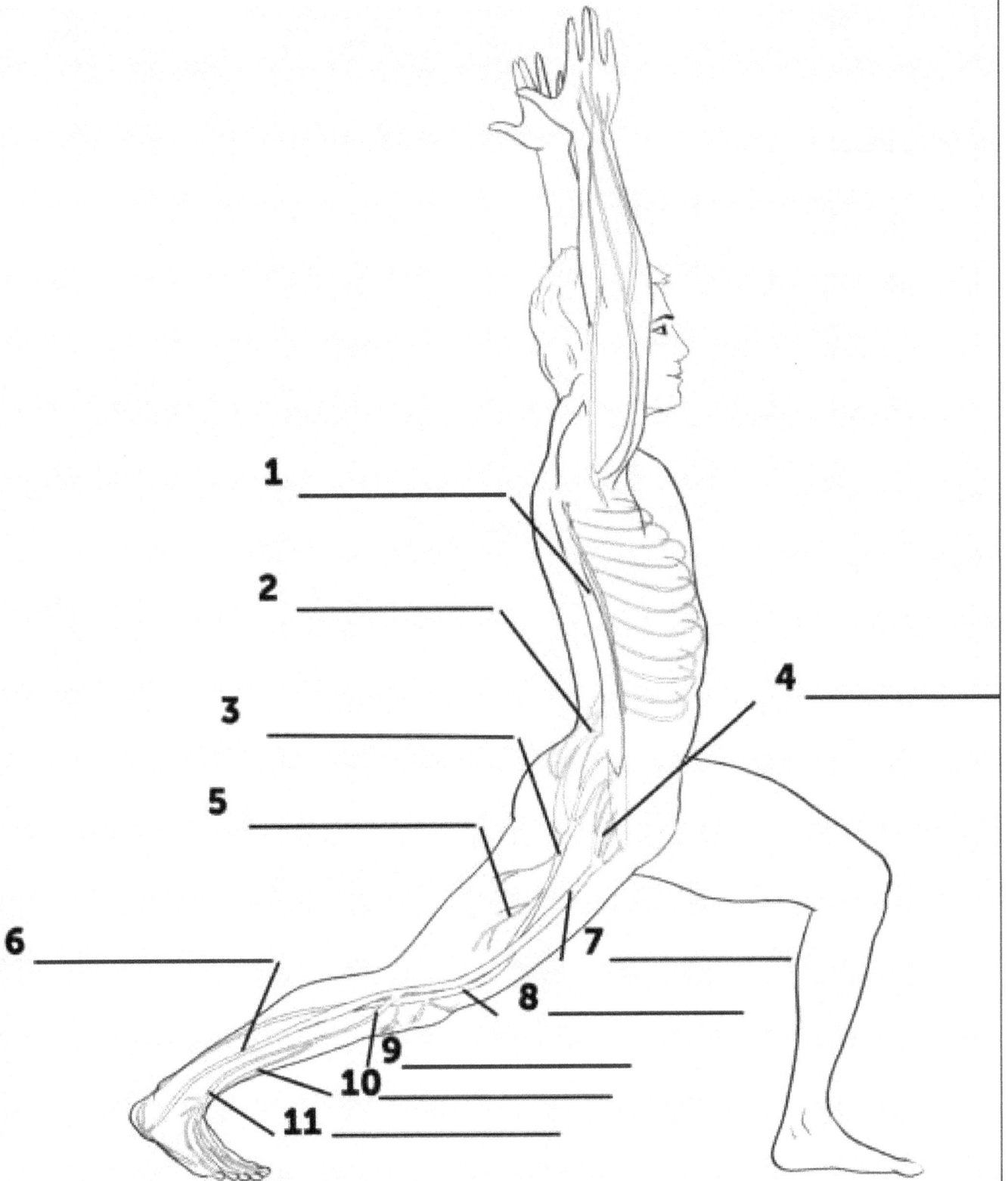

5. HIGH LUNGE

1 _____

2 _____

3 _____

4 _____

5 _____

6 _____

7 _____

8 _____

9 _____

10 _____

11 _____

5. HIGH LUNGE

1. SPINAL CORD
2. LUMBAR PLEXUS
3. FEMORAL
4. SACRAL PLEXUS
5. MUSCULAR BRANCHES OF FEMORAL
6. SCIATIC
7. SCIATIC
8. SAPHENOUS
9. COMMON PERONEAL
10. SURAL
11. SUPERFICIAL PERONEAL

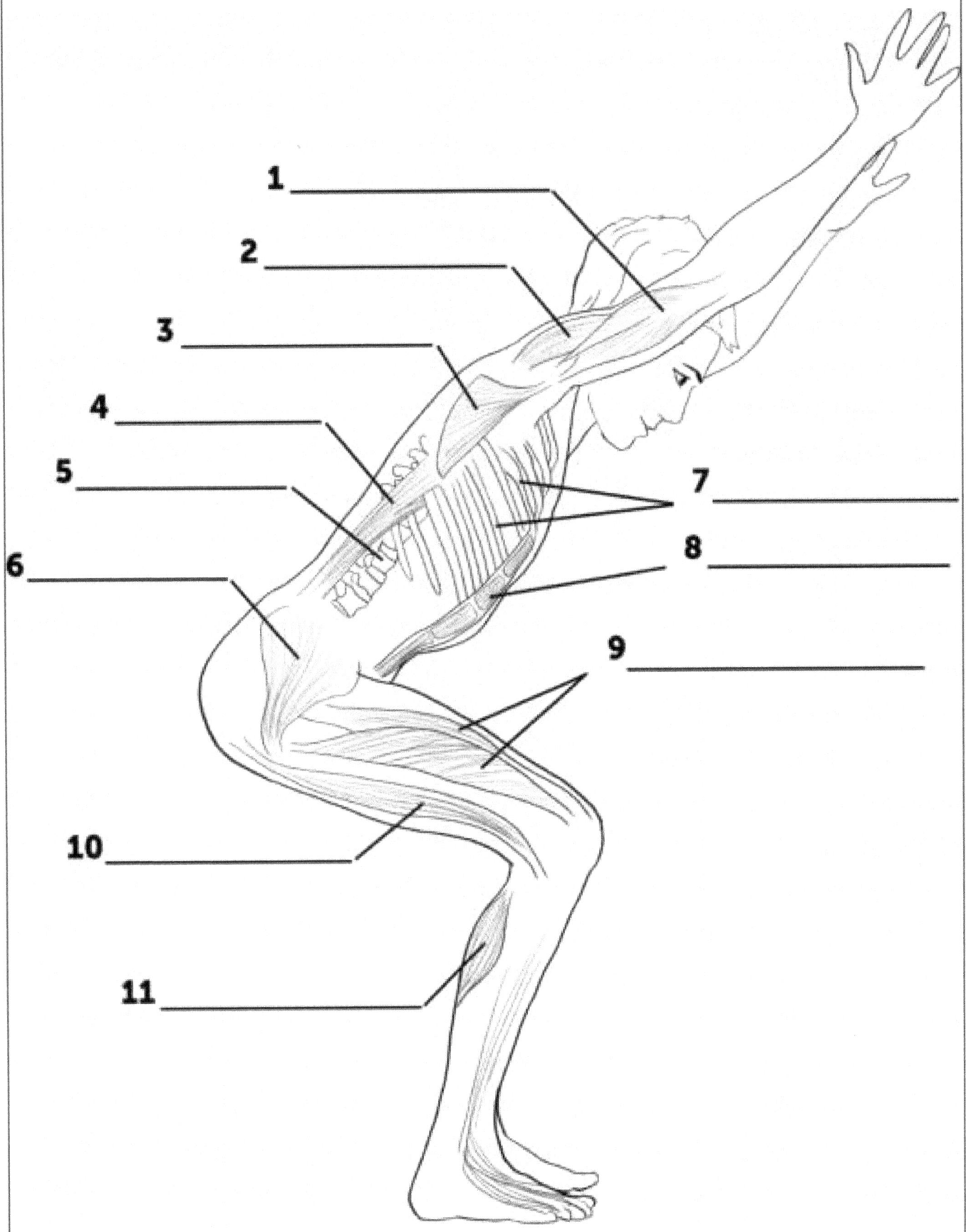

6. CHAIR POSE

1 _____

2 _____

3 _____

4 _____

5 _____

6 _____

7 _____

8 _____

9 _____

10 _____

11 _____

6. CHAIR POSE

1. TRICEPS BRACHII
2. DELTOID
3. INFRASPINATUS
4. ERECTOR SPINAE
5. SPINE
6. GLUTEUS MEDIUS
7. RIBS
8. RECTUS ABDOMINIS
9. QUADRICEPS
10. HAMSTRINGS
11. GASTROCNEMIUS

7. TRIANGLE POSE

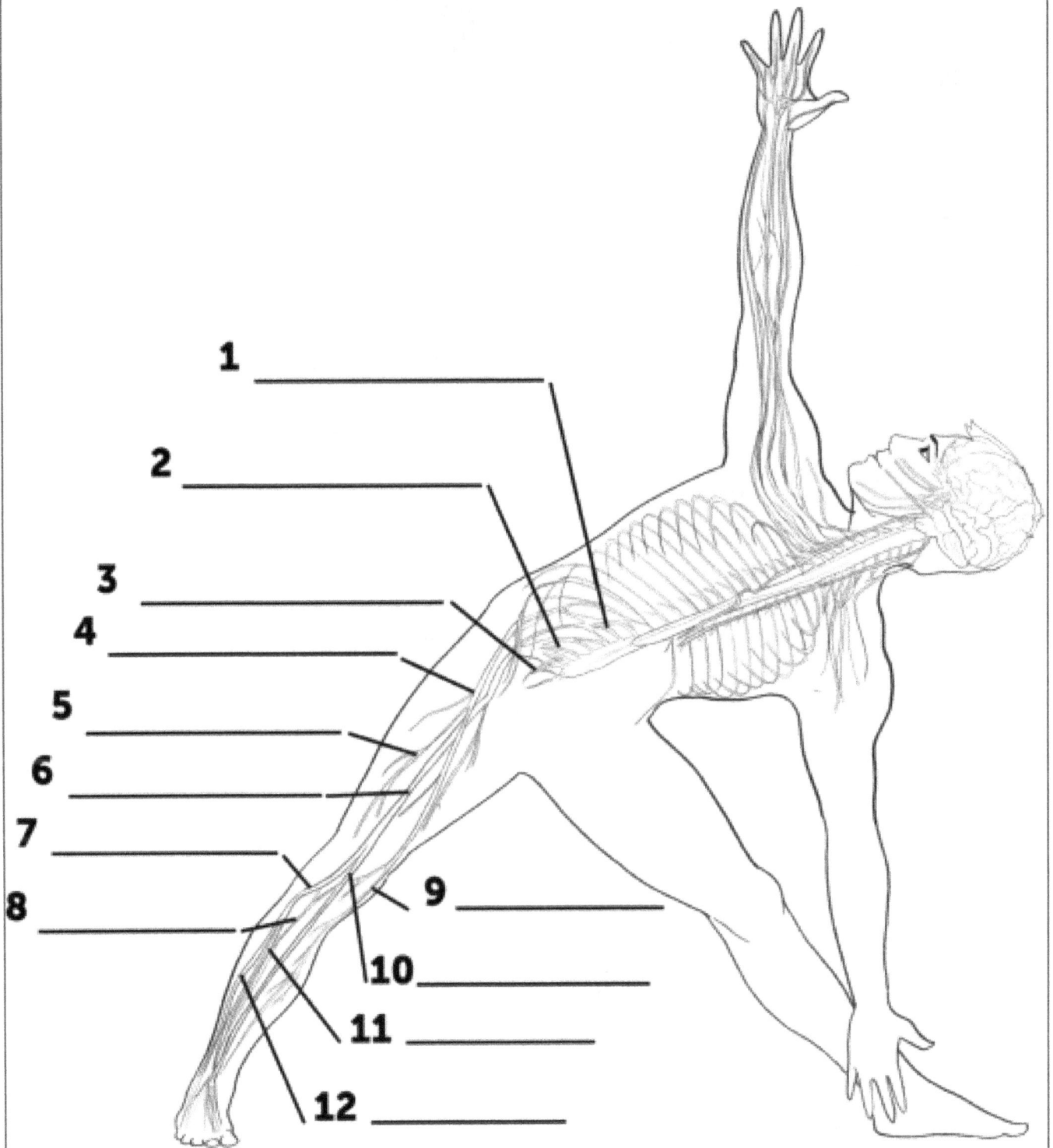

1 _____

2 _____

3 _____

4 _____

5 _____

6 _____

7 _____

8 _____

9 _____

10 _____

11 _____

12 _____

7. TRIANGLE POSE

1. LUMBAR PLEXUS
2. SACRAL PLEXUS
3. PUDENTAL NERVE
4. FEMORAL
5. MUSCULAR BRANCHES OF FEMORAL
6. SCIATIC
7. COMMON PERONEAL
8. SURAL
9. SAPHENOUS
10. TIBIAL
11. DEEP PERONEAL
12. SUPERFICIAL PERONEAL

8. EXTENDED SIDE ANGLE POSE

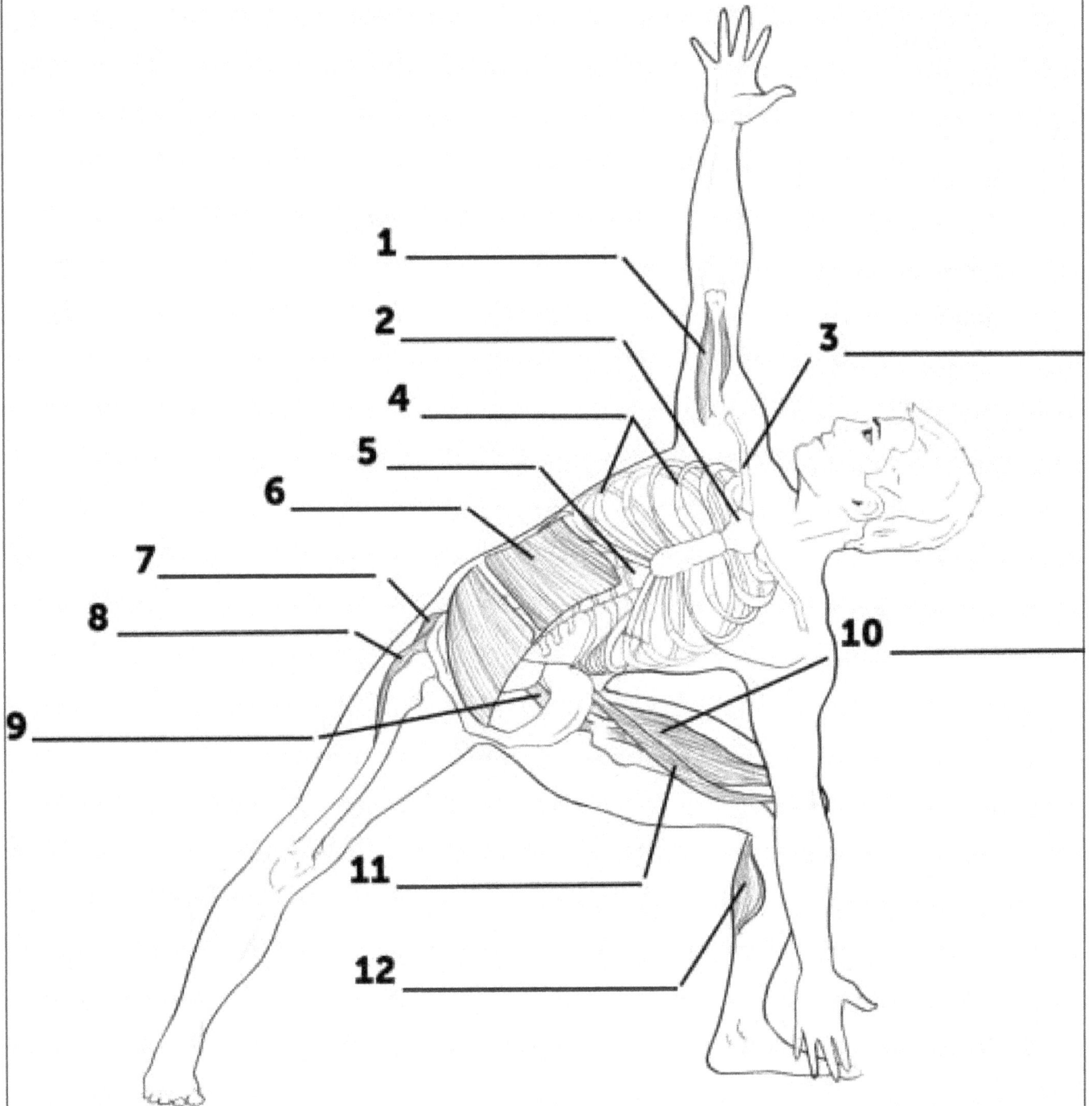

1 _____

2 _____

3 _____

4 _____

5 _____

6 _____

7 _____

8 _____

9 _____

10 _____

11 _____

12 _____

8. EXTENDED SIDE ANGLE POSE

1. BICEPS BRACHII
2. STERNUM
3. COLLARBONE
4. RIBS
5. SPINE
6. INTERNAL OBLIQUE
7. GLUTEUS MEDIUS
8. TENSOR FASCIA LATAE
9. PIRIFORMIS
10. QUADRICEPS
11. SARTORIUS
12. GASTROCNEMIUS

9. STAFF POSE

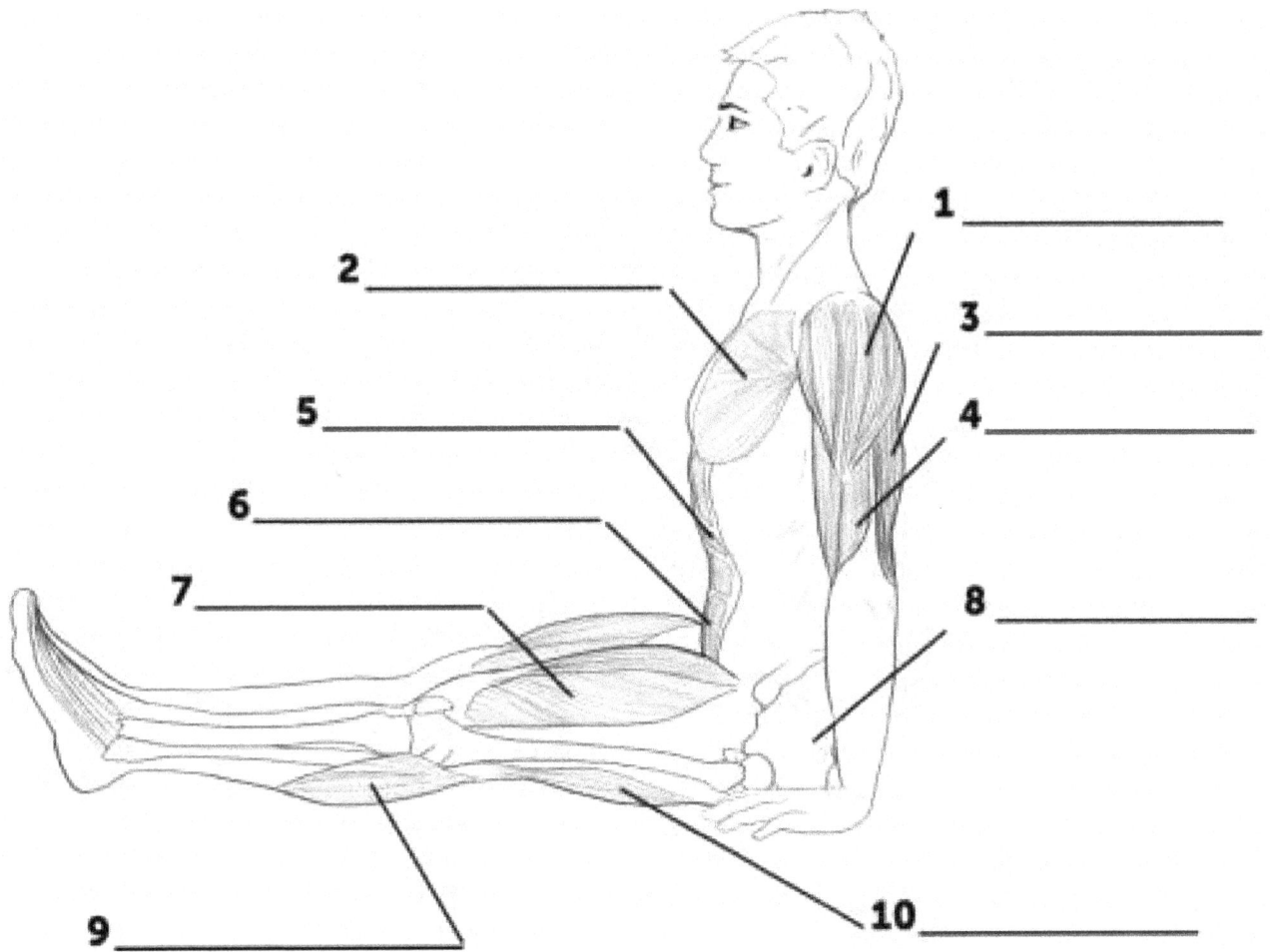

1 _____

2 _____

3 _____

4 _____

5 _____

6 _____

7 _____

8 _____

9 _____

10 _____

9. STAFF POSE

1. DELTOID
2. PECTORALIS MAJOR
3. TRICEPS BRACHII
4. BICEPS BRACHII
5. RECTUS ABDOMINIS
6. LOWER ABDOMEN MUSCLES
7. QUADRICEPS
8. PELVIS
9. GASTROCNEMIUS
10. HAMSTRINGS

10. EASY POSE

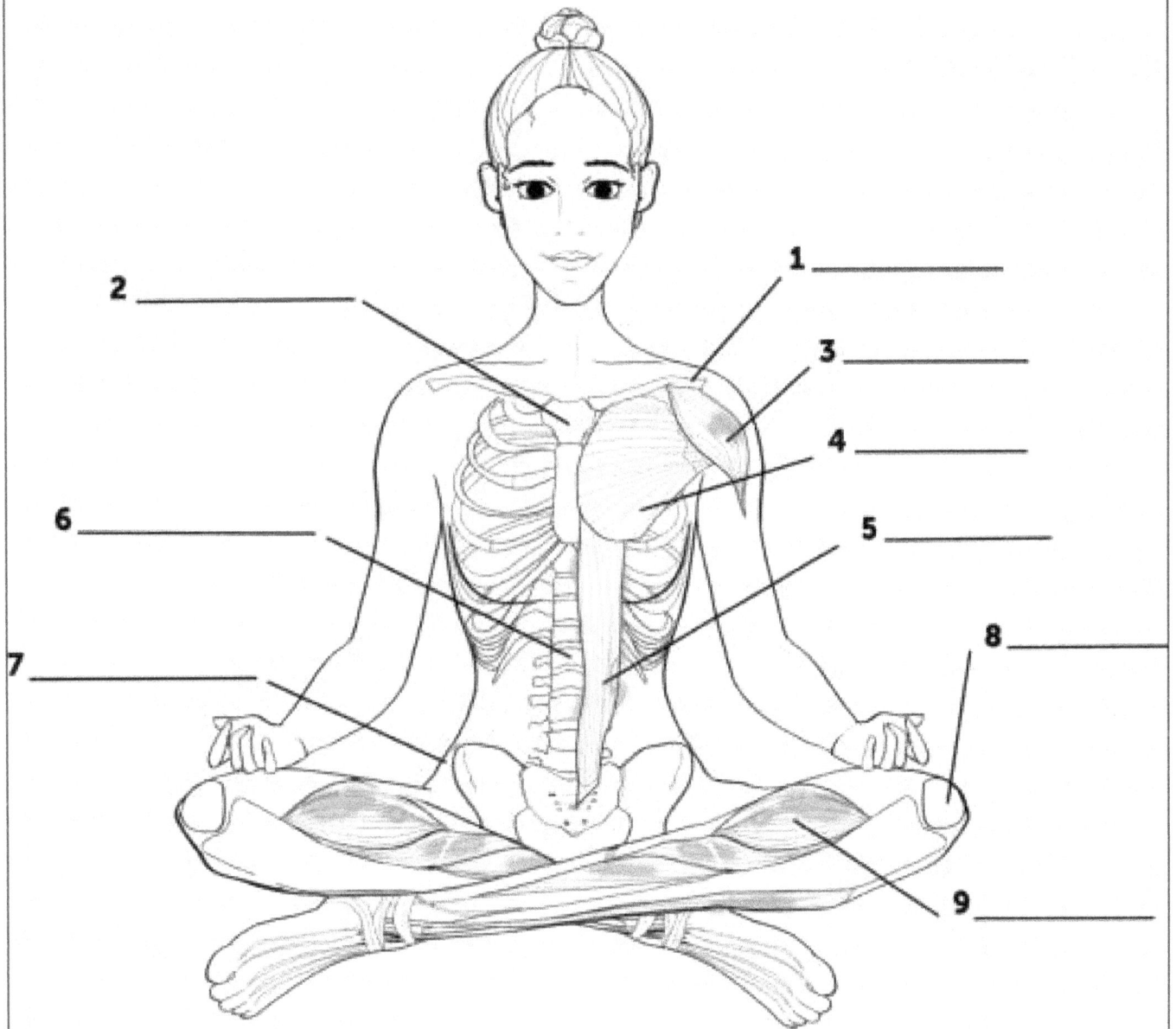

1 _____

2 _____

3 _____

4 _____

5 _____

6 _____

7 _____

8 _____

9 _____

10. EASY POSE

1. COLLARBONE
2. STERNUM
3. DELTOID
4. PECTORALIS MAJOR
5. RECTUS ABDOMINIS
6. SPINE
7. PELVIS
8. PATELLA
9. GASTROCNEMIUS

11. BOUND ANKLE

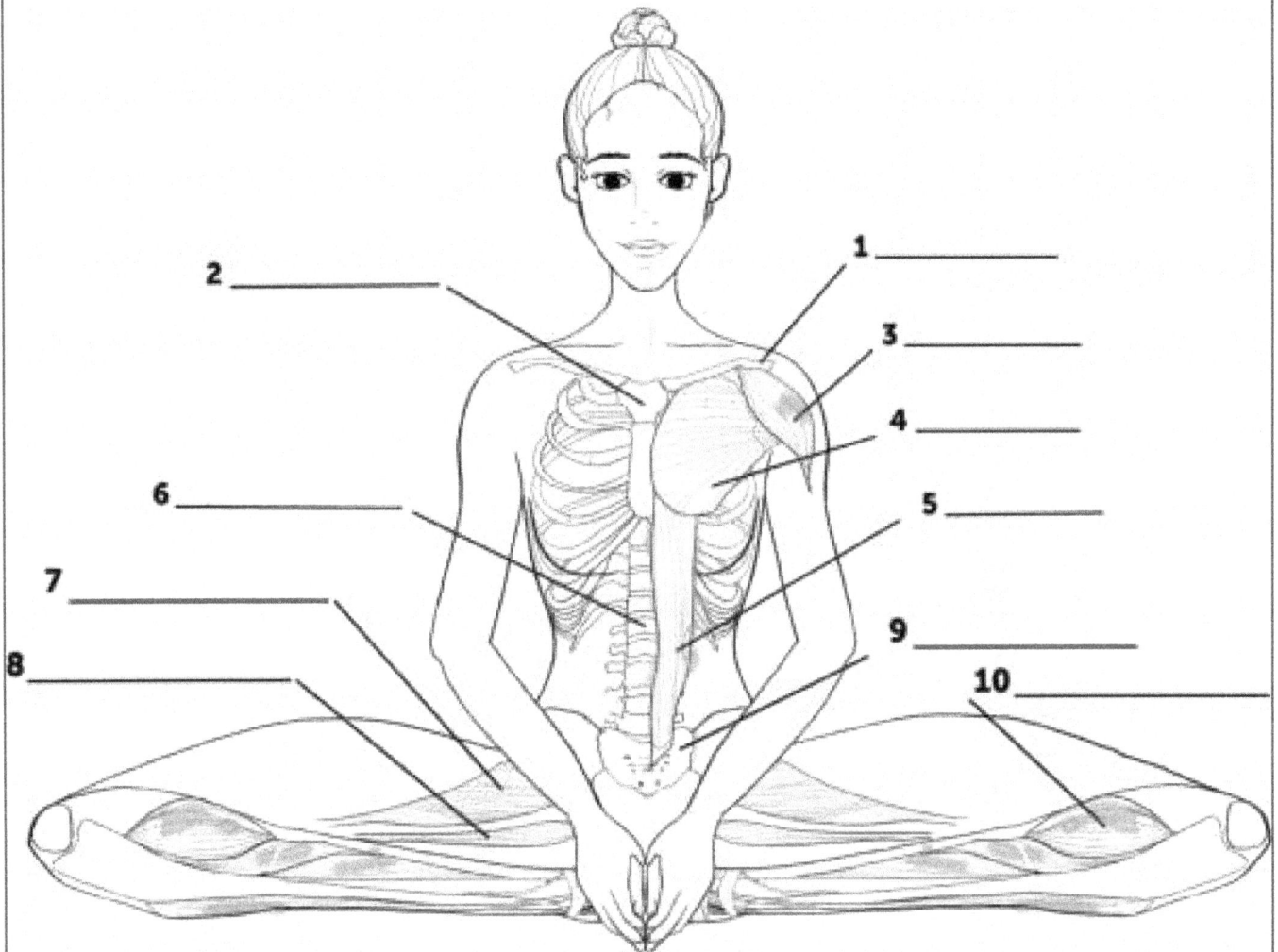

1 _____

2 _____

3 _____

4 _____

5 _____

6 _____

7 _____

8 _____

9 _____

10 _____

11. BOUND ANKLE

1. COLLARBONE
2. STERNUM
3. DELTOID
4. PECTORALIS MAJOR
5. RECTUS ABDOMINIS
6. SPINE
7. ADDUCTOR LONGUS
8. GRACILIS
9. SACRUM
10. GASTROCNEMIUS

12. HALF LORD OF THE FISHES POSE

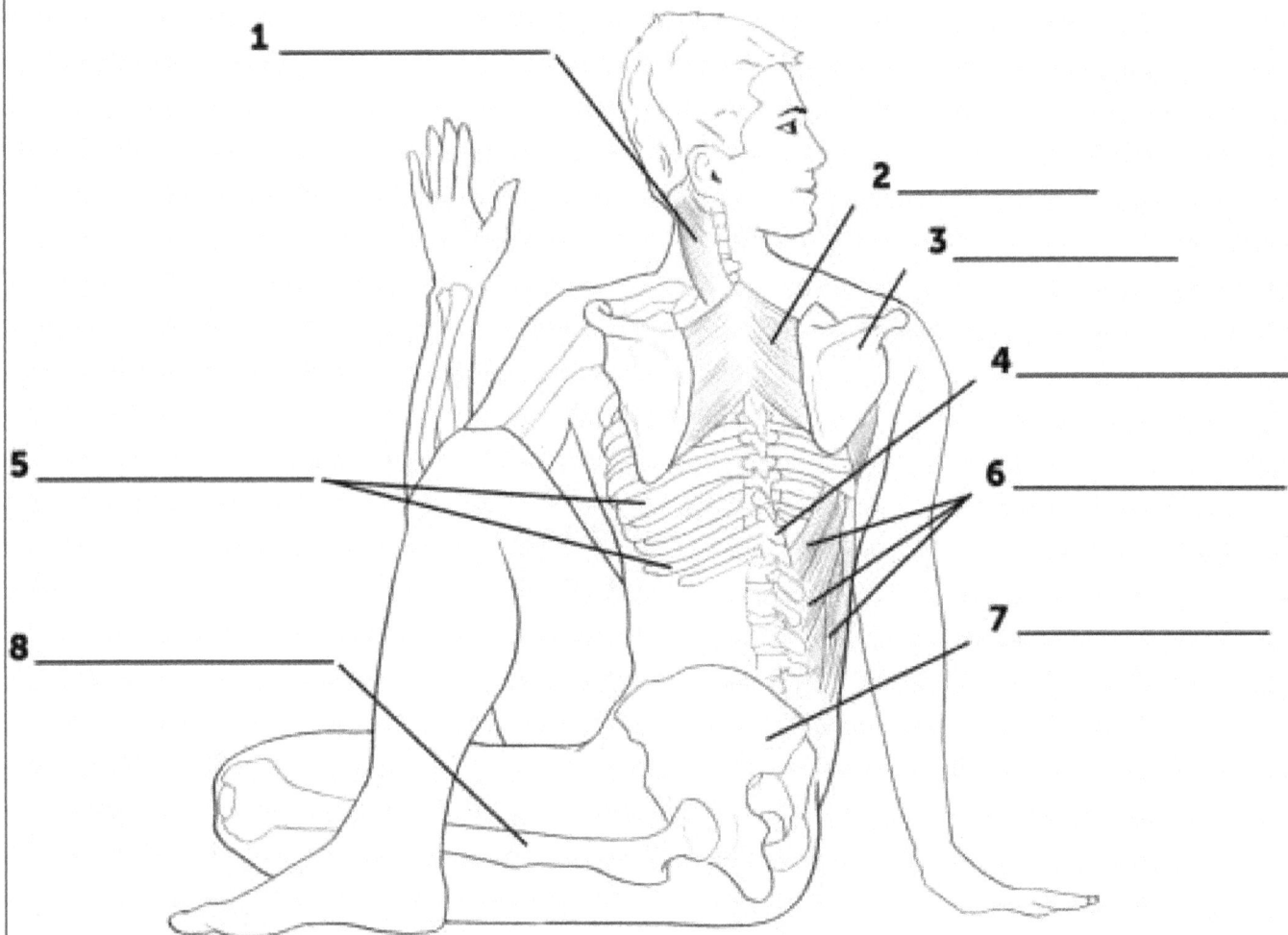

1 _____

2 _____

3 _____

4 _____

5 _____

6 _____

7 _____

8 _____

12. HALF LORD OF THE FISHES POSE

1. SPLENIUS CAPITIS
2. RHOMBOIDS
3. SCAPULA
4. SPINE
5. RIBS
6. ERECTOR SPINAE
7. PELVIS
8. FEMOR

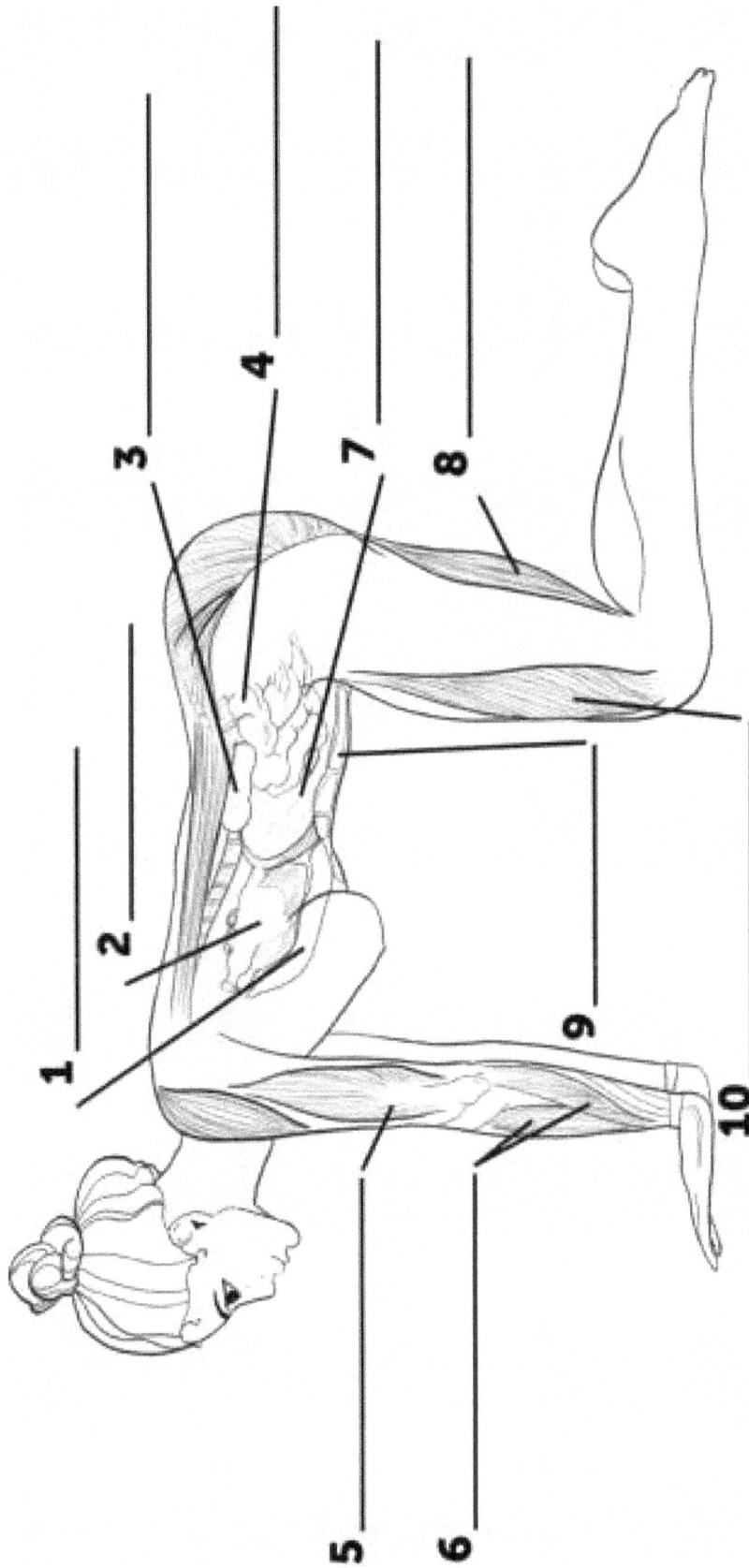

13. TABLE POSE

13. TABLE POSE

1. LUNGS
2. HEART
3. KIDNEY
4. ASCENDING COLON
5. TRICEPS BRACHII
6. PRONATORS
7. LIVER
8. HAMSTRINGS
9. RECTUS ABDOMINIS
10. QUADRICEPS

14. CAT POSE

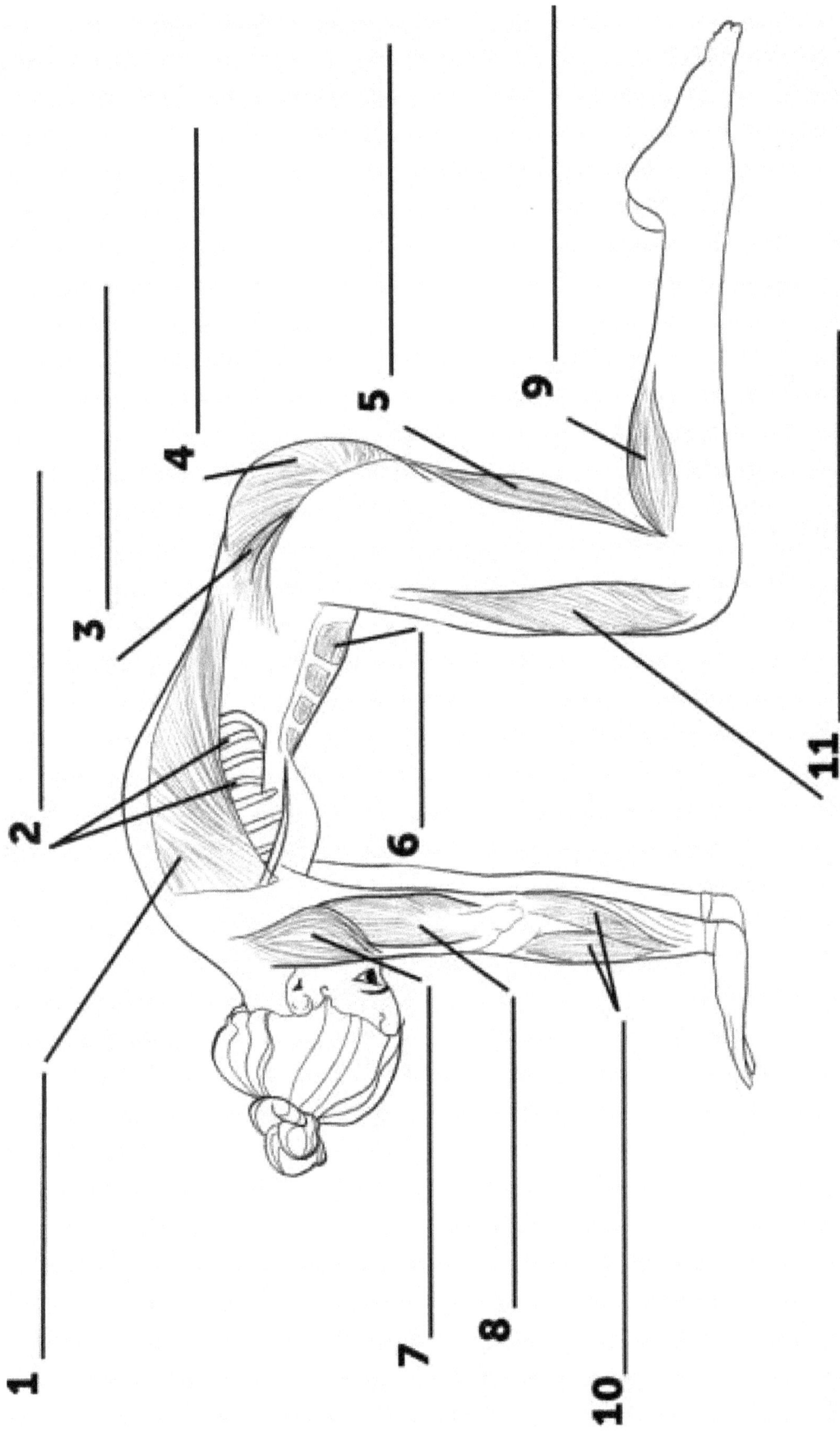

14. CAT POSE

1. LATISSIMUS DORSI
2. RIBS
3. PIRIFORMIS
4. GLUTEUS MAXIMUS
5. HAMSTRINGS
6. RECTUS ABDOMINIS
7. DELTOID
8. TRICEPS BRACHII
9. GASTROCNEMIUS
10. PRONATORS
11. QUADRICEPS

15. COW POSE

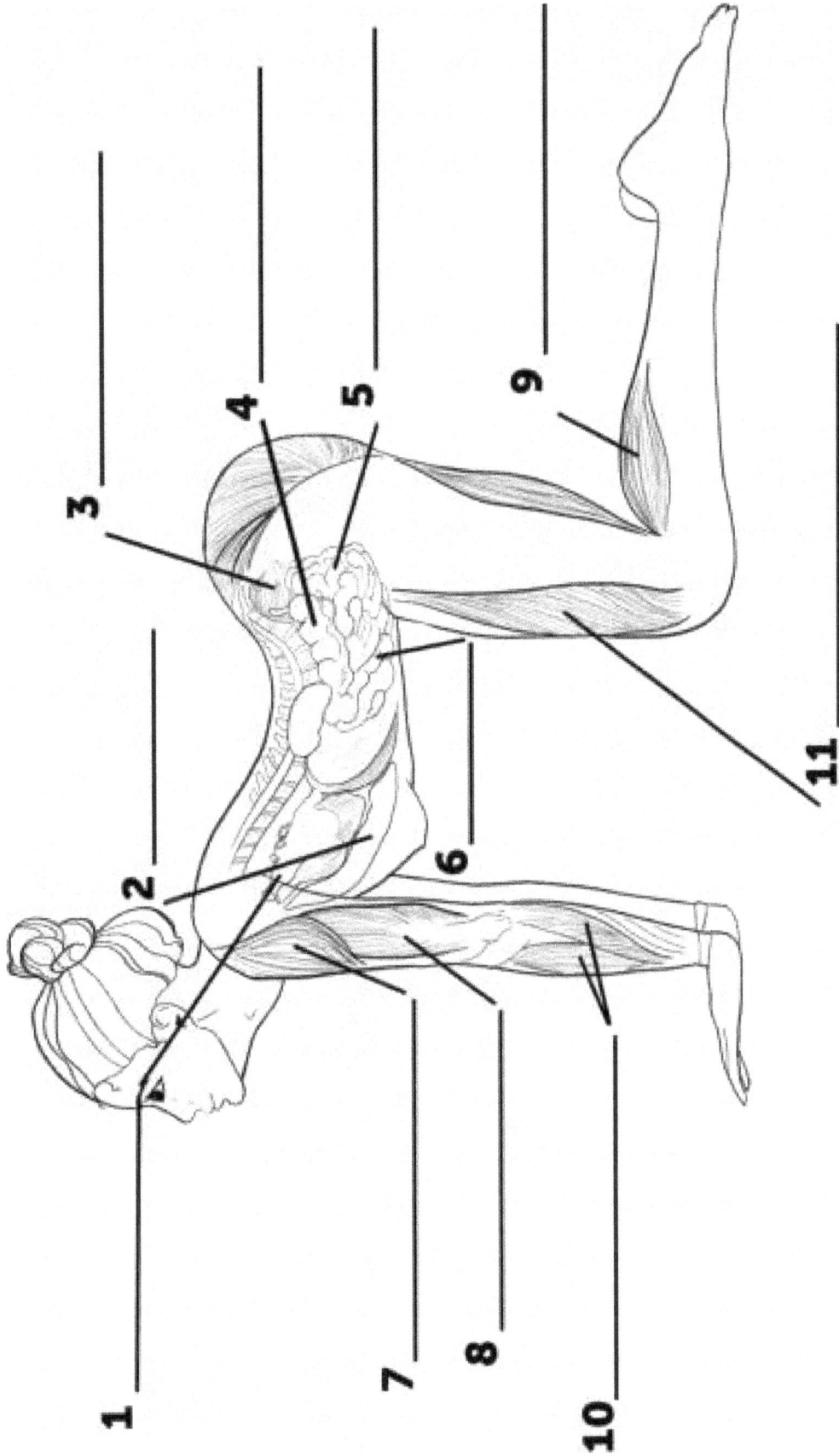

15. COW POSE

1. HEART
2. LUNGS
3. RECTUM
4. ASCENDING COLON
5. COILS OF SMALL INTESTINE
6. TRANSVERSE COLON
7. DELTOID
8. TRICEPS BRACHII
9. GASTROCNEMIUS
10. PRONATORS
11. QUADRICEPS

16. BALANCING TABLE POSE

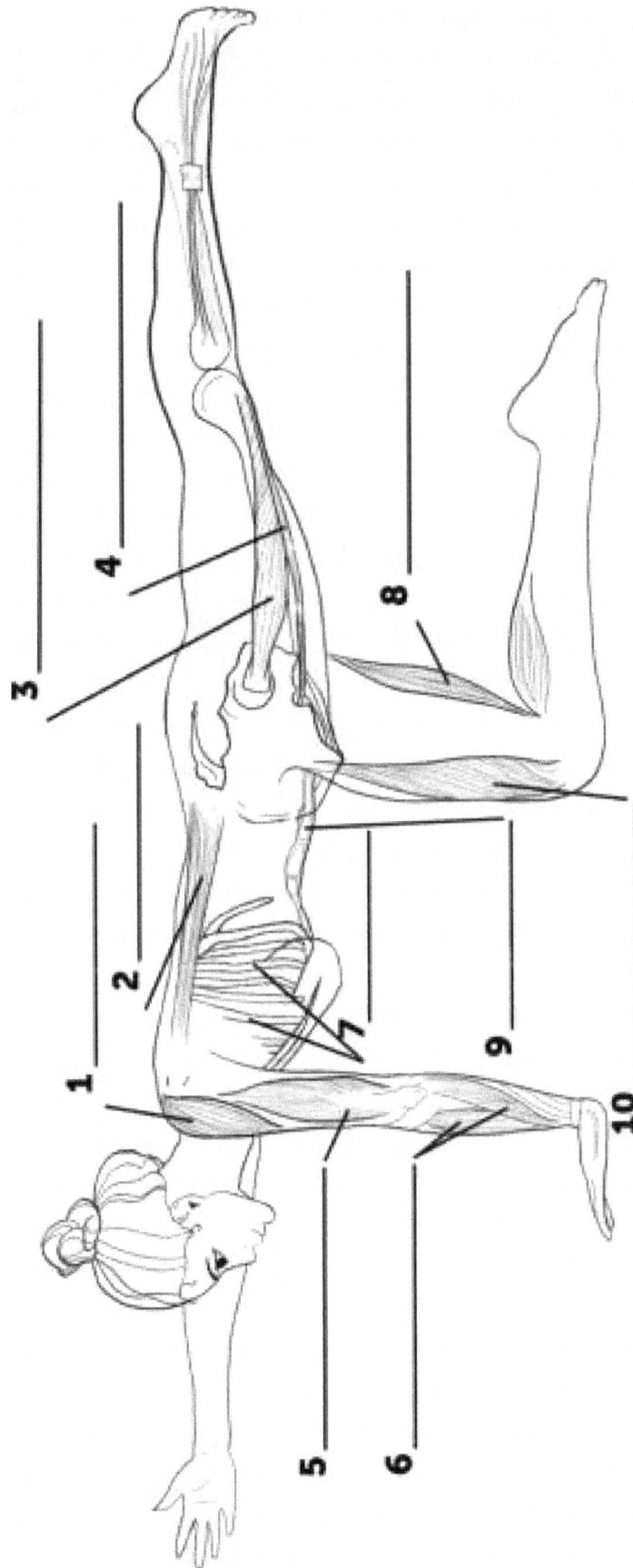

16. BALANCING TABLE POSE

1.   DELTOID

2.   ERECTOR SPINAE

3.   RECTUS FEMORIS

4.   SARTORIUS

5.   TRICEPS BRACHII

6.   PRONATORS

7.   RIBS

8.   HAMSTRINGS

9.   RECTUS ABDOMINIS

10. QUADRICEPS

17. REVERSE TABLE TOP POSE

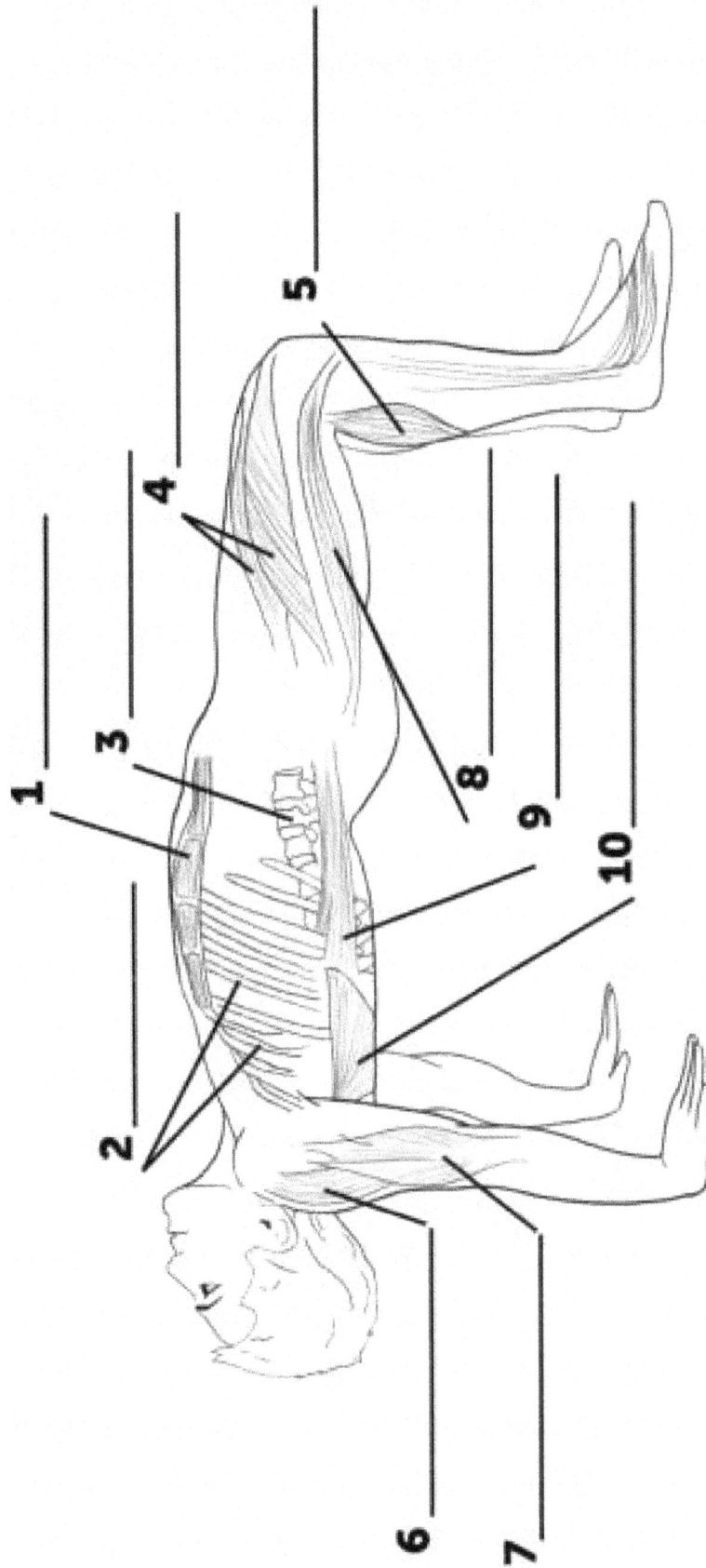

17. REVERSE TABLE TOP POSE

1. RECTUS ABDOMINIS

2. RIBS

3. SPINE

4. QUADRICEPS

5. GASTROCNEMIUS

6. DELTOID

7. TRICEPS BRACHII

8. HAMSTRINGS

9. ERECTOR SPINAE

10. INFRASPINATUS

18. SPHINX POSE

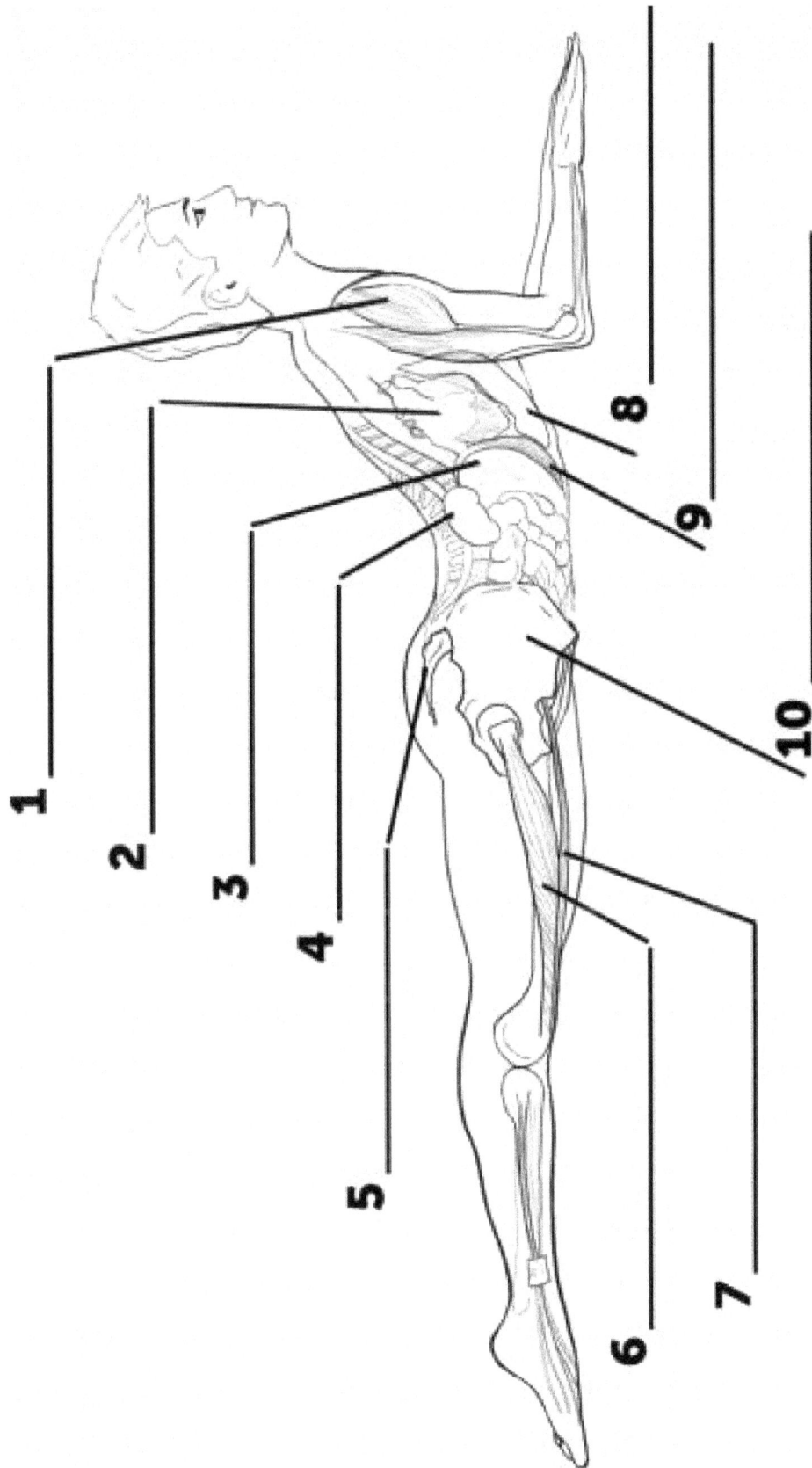

1

2

3

4

5

6

7

8

9

10

18. SPHINX POSE

1. DELTOID
2. HEART
3. LIVER
4. KIDNEY
5. SACRUM
6. RECTUS FEMORIS
7. SARTORIUS
8. LUNGS
9. DIAPHRAGM
10. PELVIS

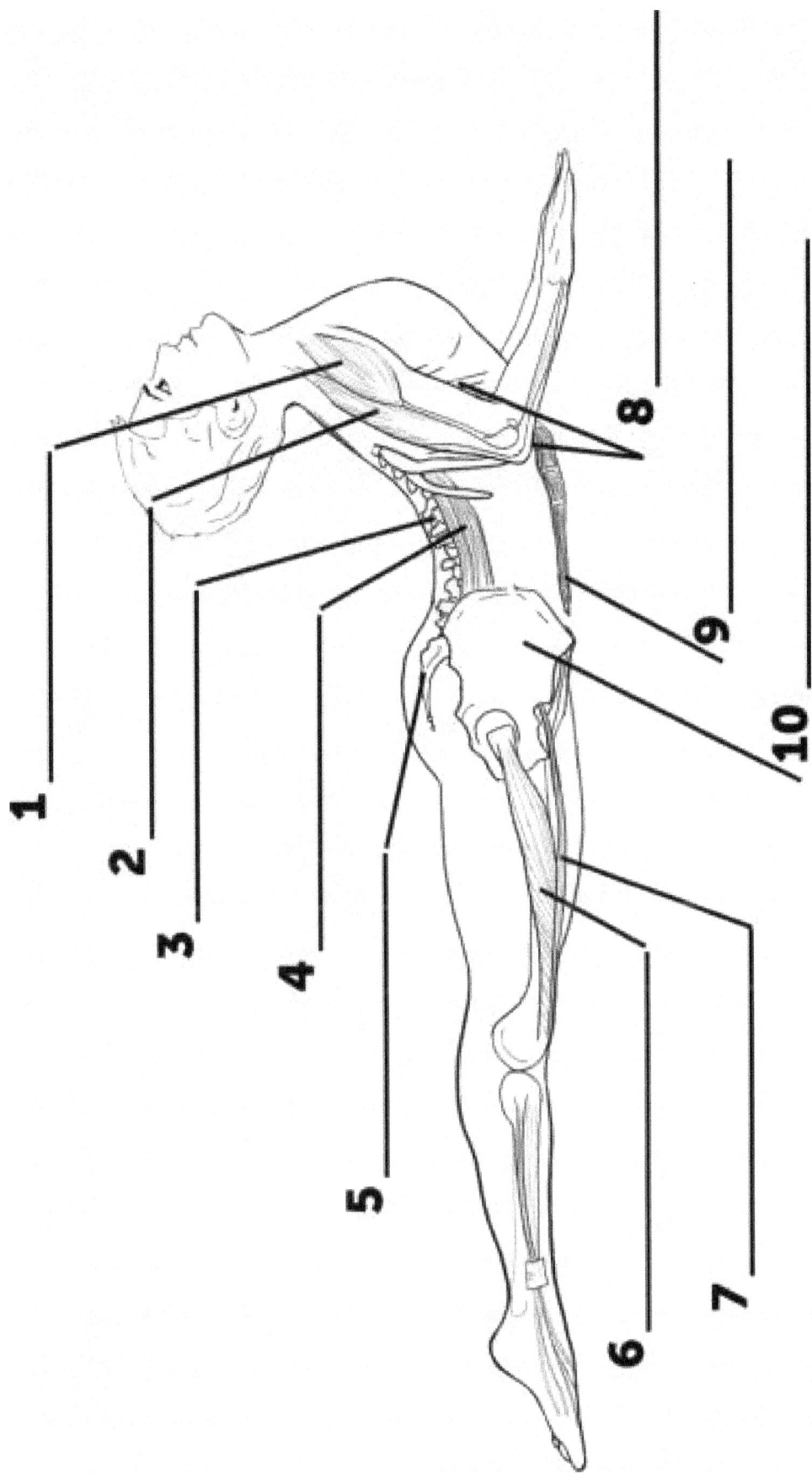

19. COBRA POSE

1

2

3

4

5

6

7

8

9

10

19. COBRA POSE

1. DELTOID
2. TRICEPS BRACHII
3. SPINE
4. ERECTOR SPINAE
5. SACRUM
6. RECTUS FEMORIS
7. SARTORIUS
8. RIBS
9. RECTUS ABDOMINIS
10. PELVIS

20. BIG TOE POSE

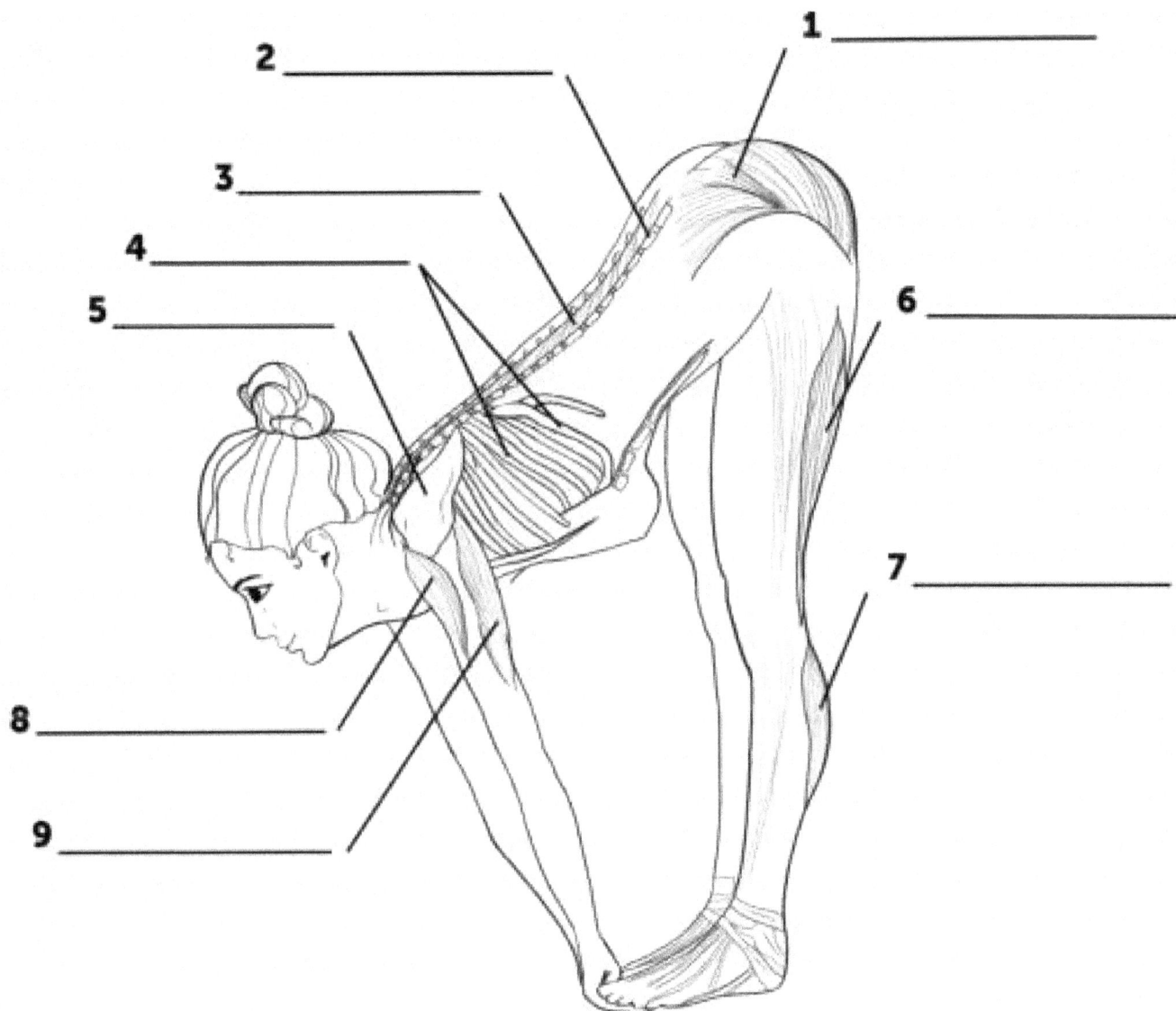

1 _____

2 _____

3 _____

4 _____

5 _____

6 _____

7 _____

8 _____

9 _____

20. BIG TOE POSE

1. PIRIFORMIS
2. SPINE
3. SPINAL MUSCLES
4. RIBS
5. SCAPULA
6. HAMSTRINGS
7. GASTROCNEMIUS
8. DELTOID
9. TRICEPS BRACHII

21. CHILD'S POSE

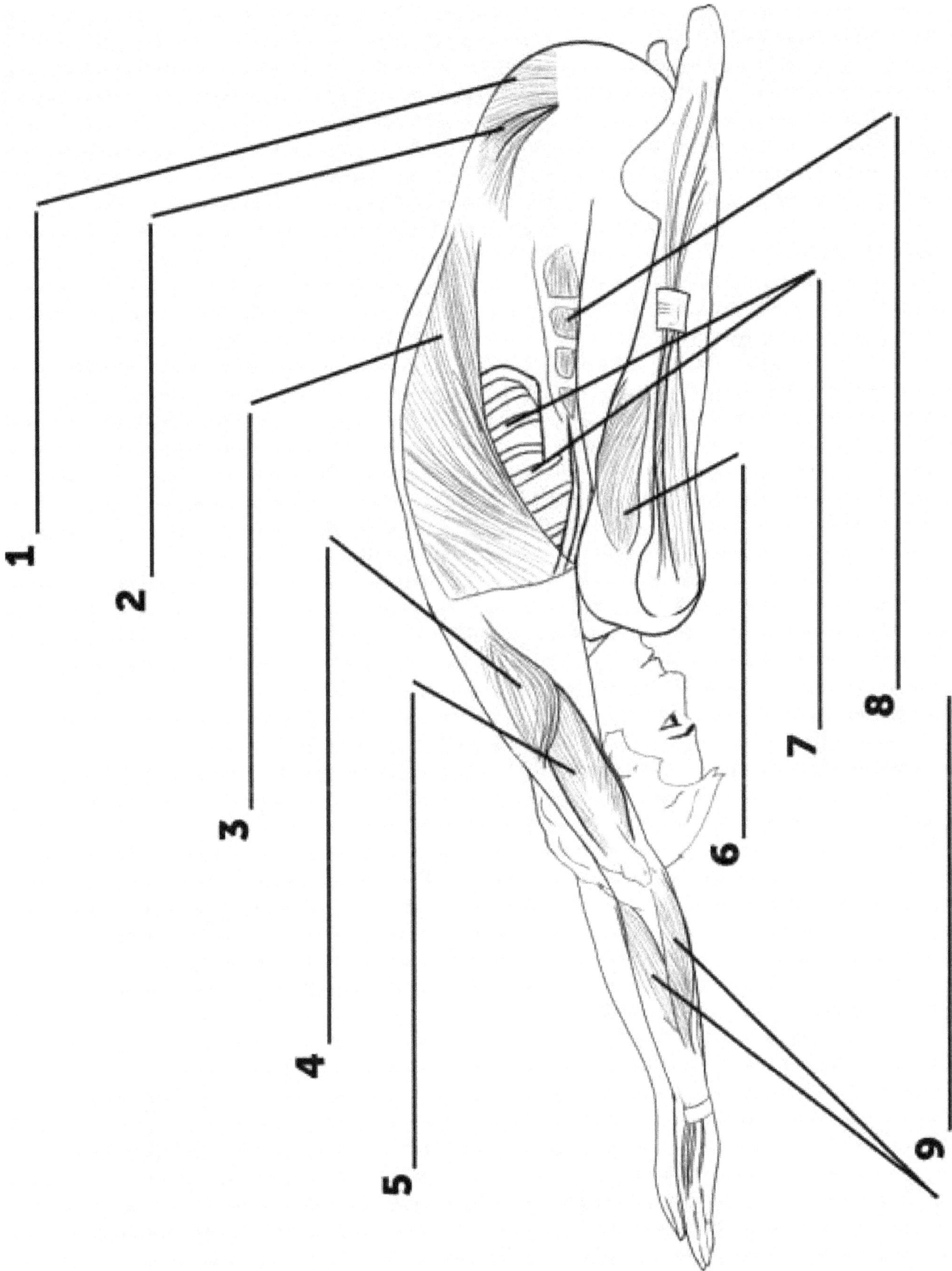

1

2

3

4

5

6

7

8

9

21. CHILD'S POSE

1. GLUTEUS MAXIMUS
2. PIRIFORMIS
3. LATISSIMUS DORSI
4. DELTOID
5. TRICEPS BRACHII
6. GASTROCNEMIUS
7. RIBS
8. RECTUS ABDOMINIS
9. PRONATORS

22. ONE-LEGGED BOAT POSE

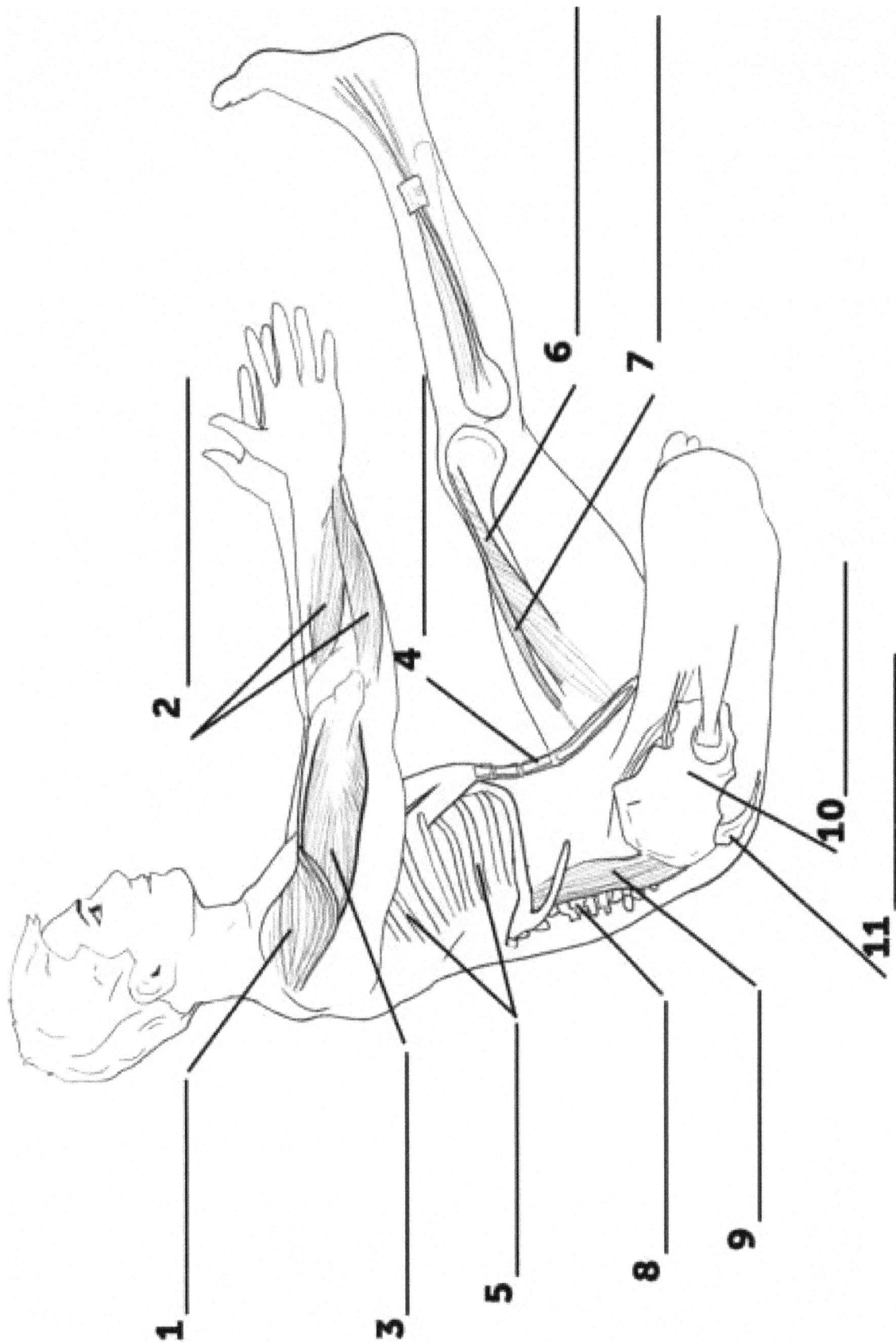

22. ONE-LEGGED BOAT POSE

1. DELTOID
2. PRONATORS
3. TRICEPS BRACHII
4. RECTUS ABDOMINIS
5. RIBS
6. RECTUS FEMORIS
7. SARTORIUS
8. SPINE
9. ERECTOR SPINAE
10. PELVIS
11. SACRUM

23. DOLPHIN POSE

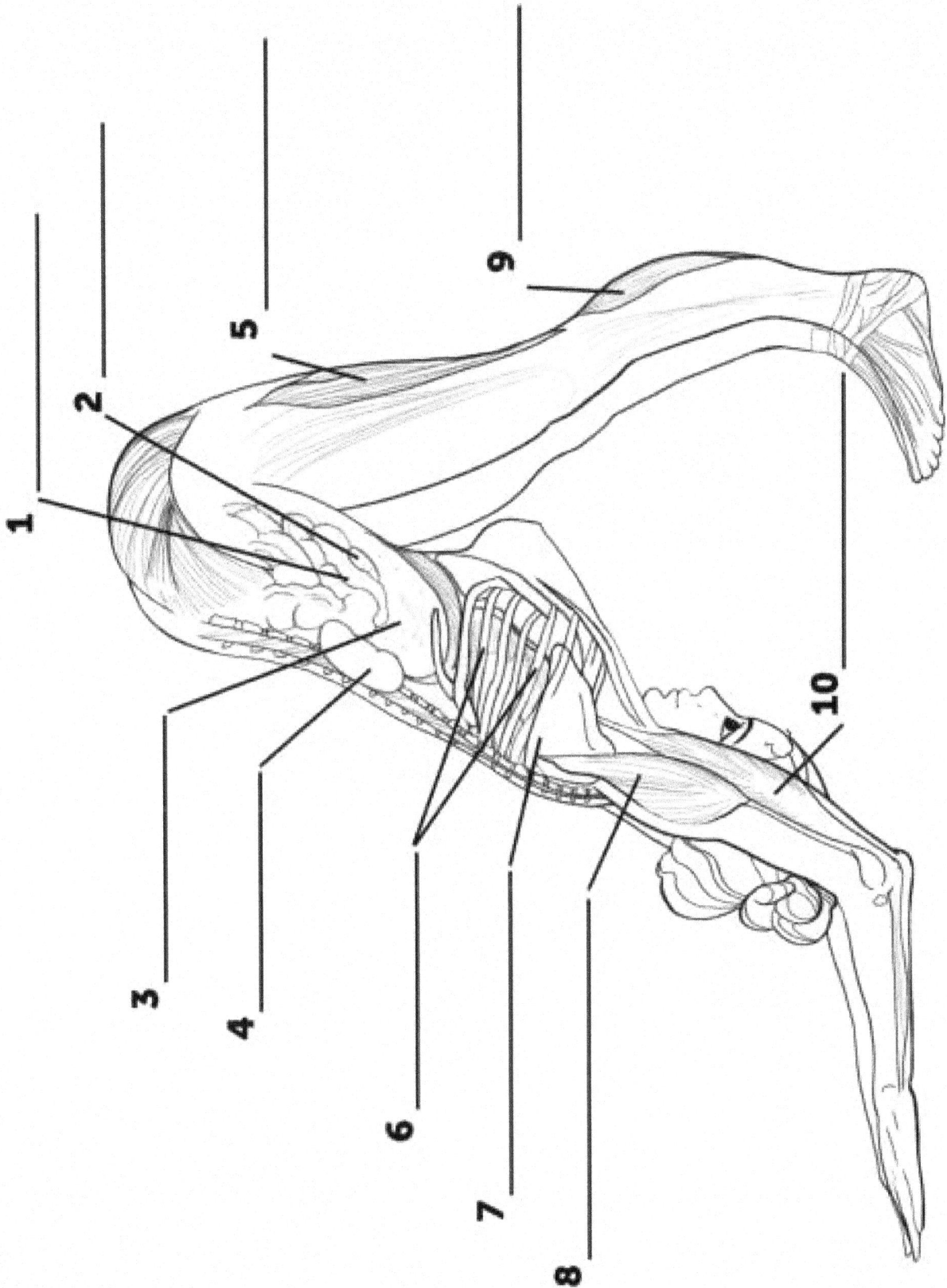

23. DOLPHIN POSE

1.   STOMACH
2.   GALLBLADDER
3.   LIVER
4.   KIDNEY
5.   HAMSTRINGS
6.   RIBS
7.   SCAPULA
8.   DELTOID
9.   GASTROCNEMIUS
10. TRICEPS BRACHII

24. BRIDGE POSE

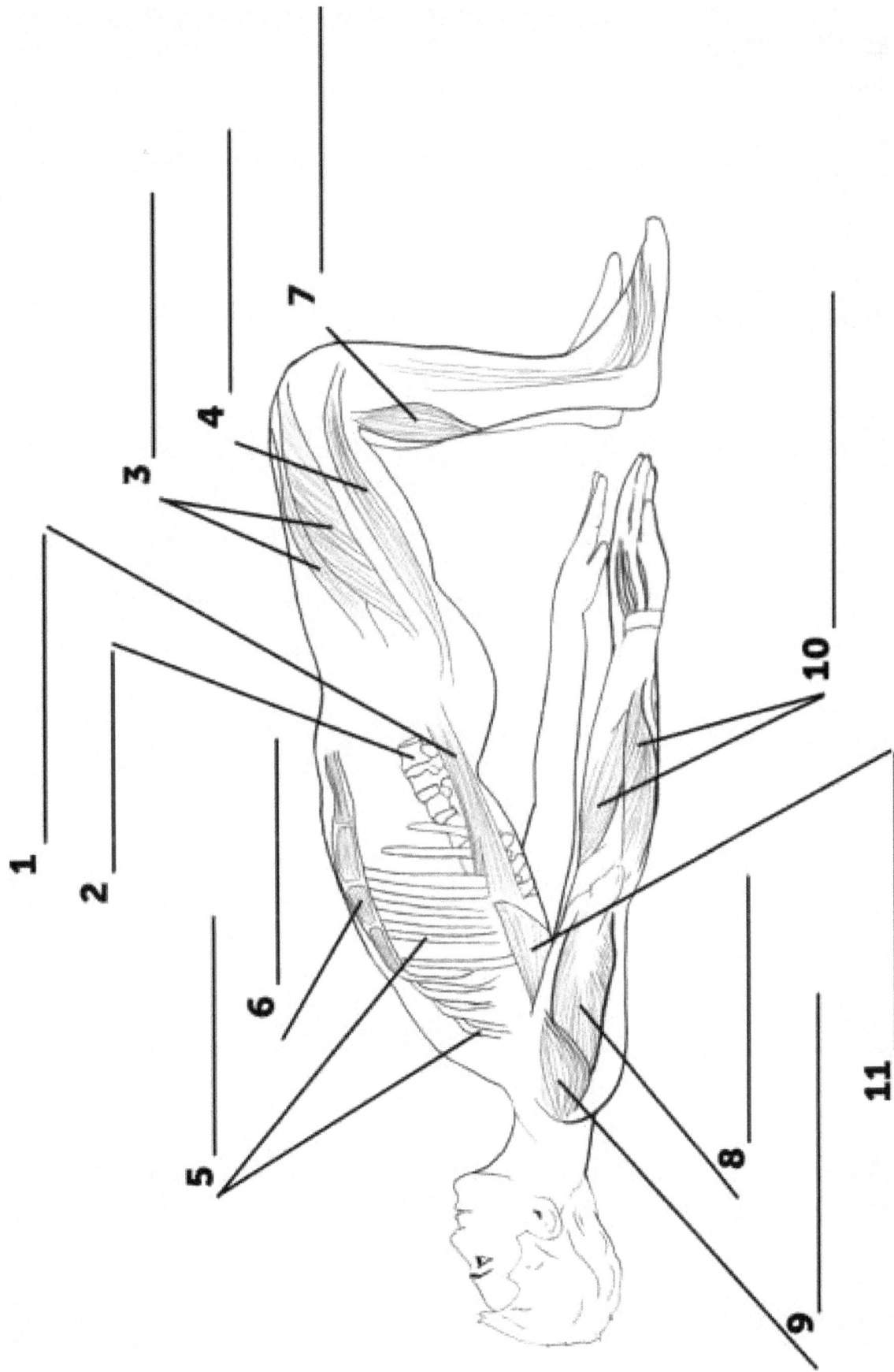

1

2

3

4

5

6

7

8

9

10

11

24. BRIDGE POSE

1. ERECTOR SPINAE
2. SPINE
3. QUADRICEPS
4. HAMSTRINGS
5. RIBS
6. RECTUS ABDOMINIS
7. GASTROCNEMIUS
8. TRICEPS BRACHII
9. DELTOID
10. PRONATORS
11. INFRASPINATUS

25. GARLAND POSE

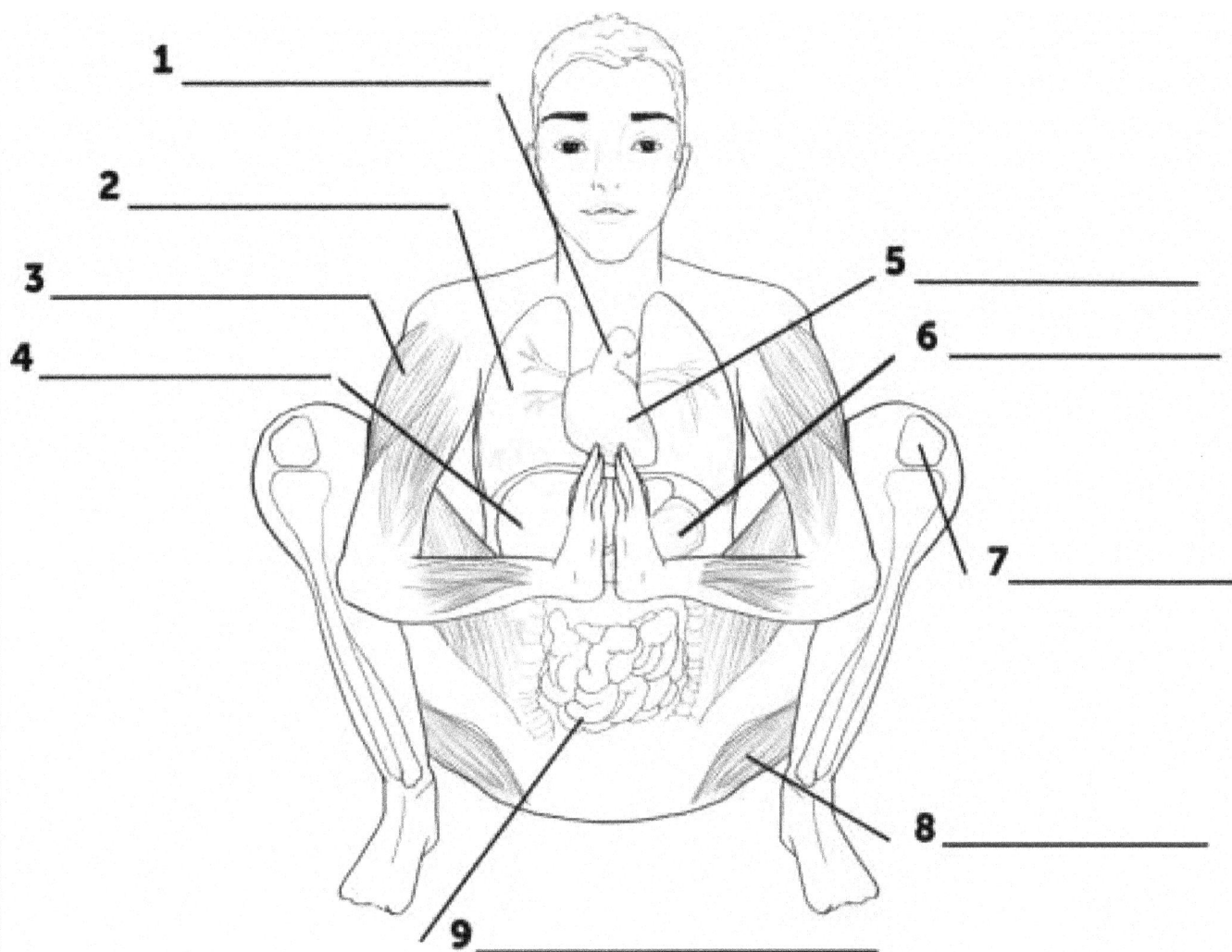

1 _____

2 _____

3 _____

4 _____

5 _____

6 _____

7 _____

8 _____

9 _____

25. GARLAND POSE

1. AORTA
2. LUNGS
3. TRICEPS BRACHII
4. LIVER
5. HEART
6. STOMACH
7. PATELLA
8. HAMSTRINGS
9. COILS OF SMALL INTESTINE

26. DOWNWARD-FACING DOG

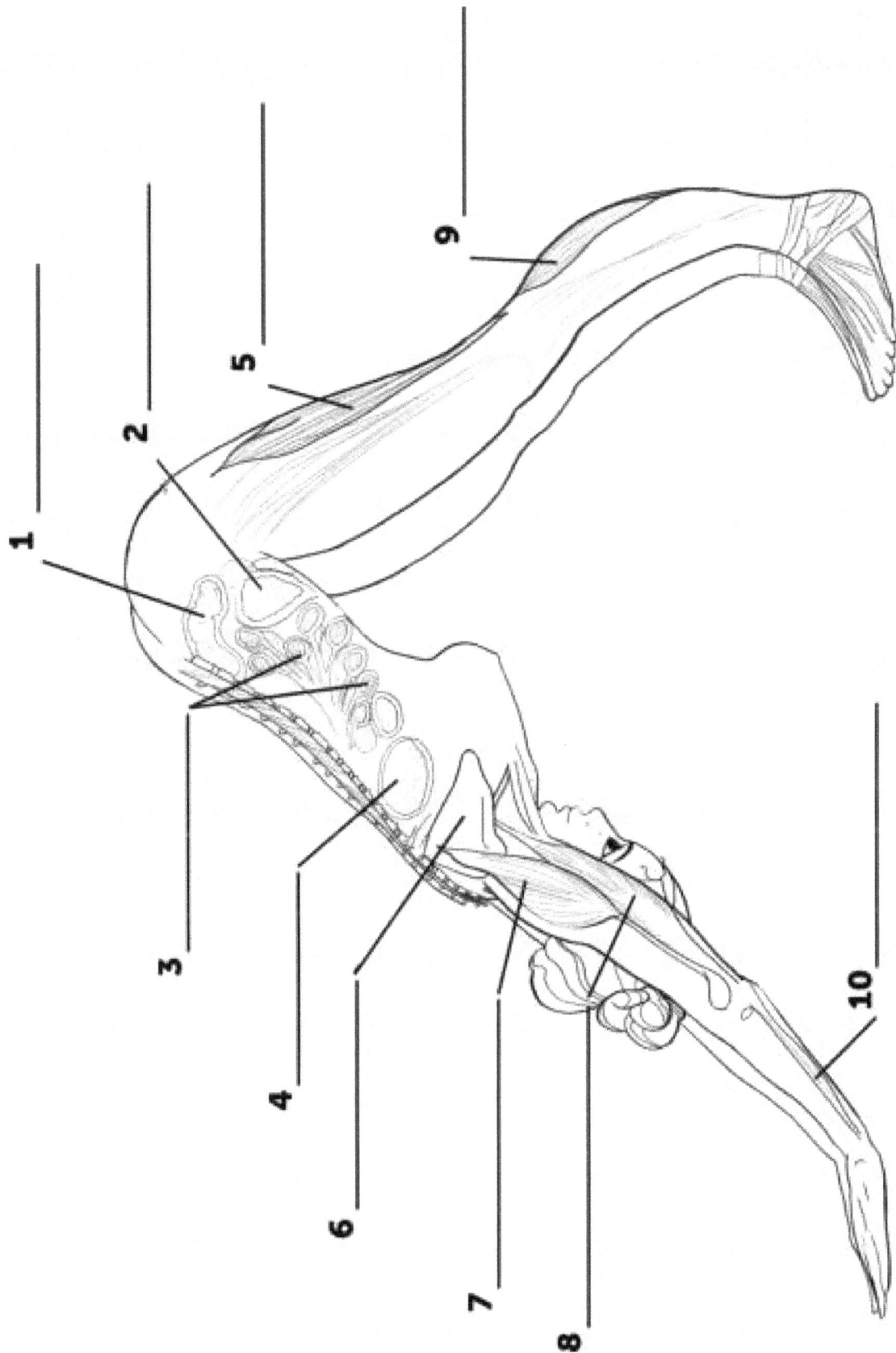

26. DOWNWARD-FACING DOG

1. RECTUM
2. URINARY BLADDER
3. SMALL INTESTINE
4. STOMACH
5. HAMSTRINGS
6. SCAPULA
7. DELTOID
8. TRICEPS BRACHII
9. GASTROCNEMIUS
10. PRONATORS

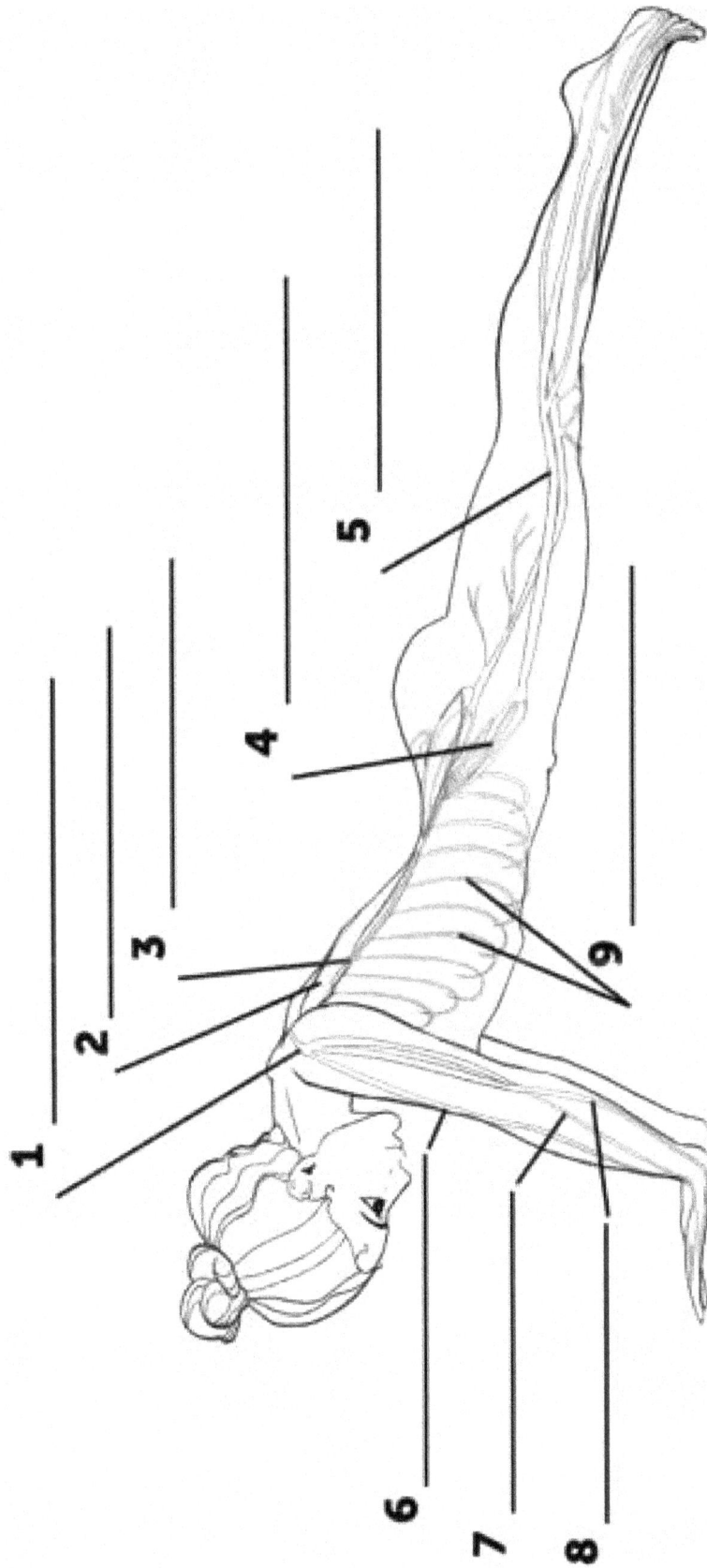

27. PLANK POSE

1

2

3

4

5

6

7

8

9

27. PLANK POSE

1. BRACHIAL PLEXUS
2. SPINAL CORD
3. VAGUS
4. LUMBAR PLEXUS
5. SCIATIC
6. ULNAR
7. MEDIAN
8. RADIAL
9. INTERCOSTALS

28. CHATURANGA

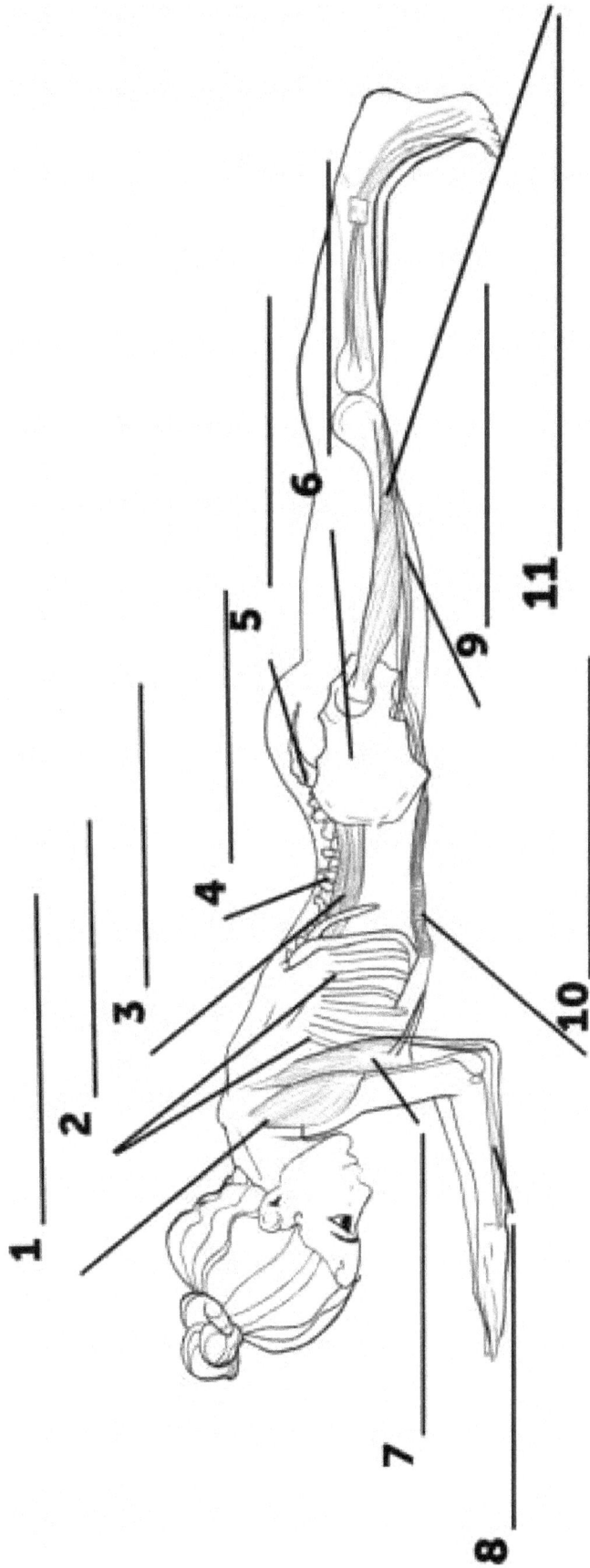

28. CHATURANGA

1. DELTOID
2. RIBS
3. ERECTOR SPINAE
4. SPINE
5. SACRUM
6. PELVIS
7. TRICEPS BRACHII
8. PRONATORS
9. SARTORIUS
10. RECTUS ABDOMINIS
11. RECTUS FEMORIS

29. UPWARD-FACING DOG

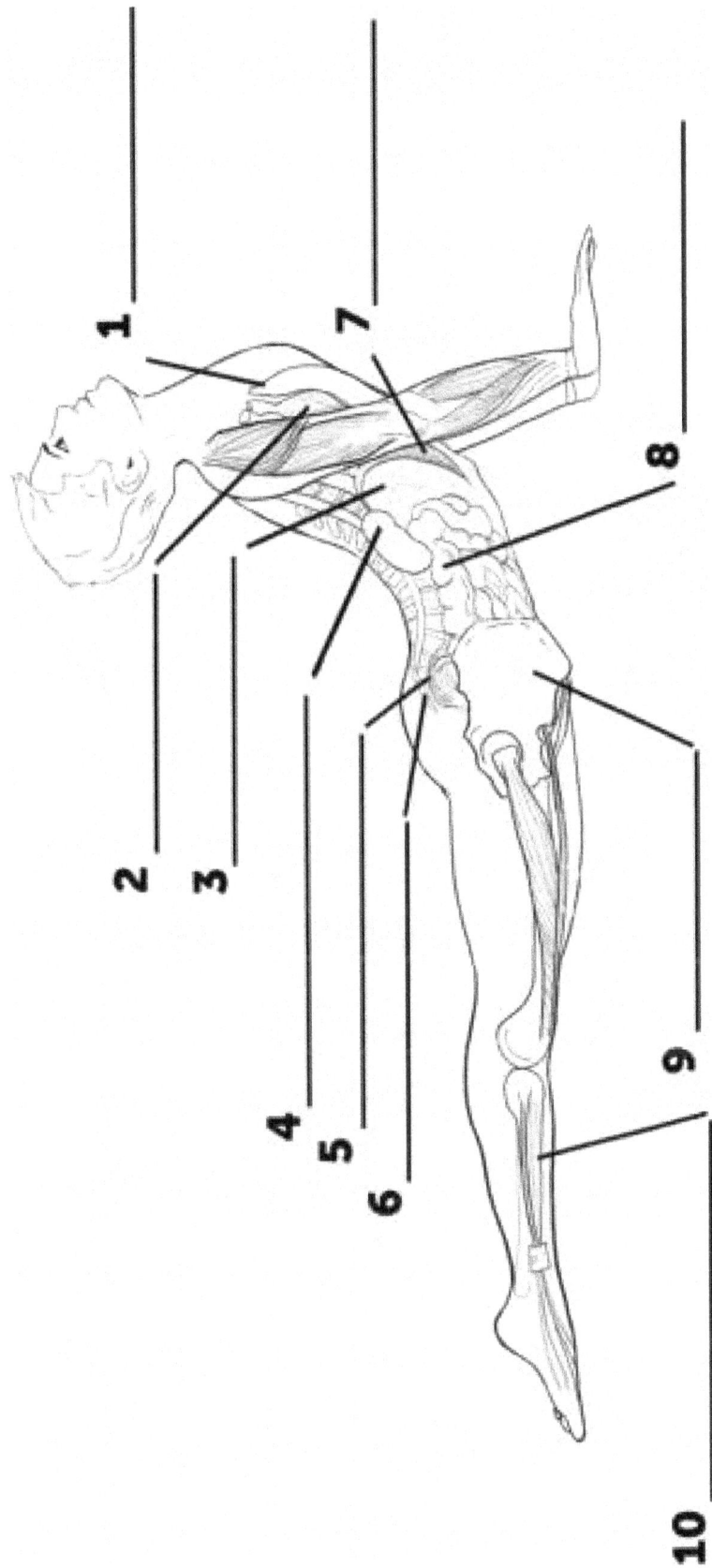

1

2

3

4

5

6

7

8

9

10

29. UPWARD-FACING DOG

1. LUNGS
2. HEART
3. LIVER
4. KIDNEY
5. RECTUM
6. SACRUM
7. DIAPHRAGM
8. ASCENDING COLON
9. PELVIS
10. TIBIALIS ANTERIOR

30. WIND REMOVING POSE

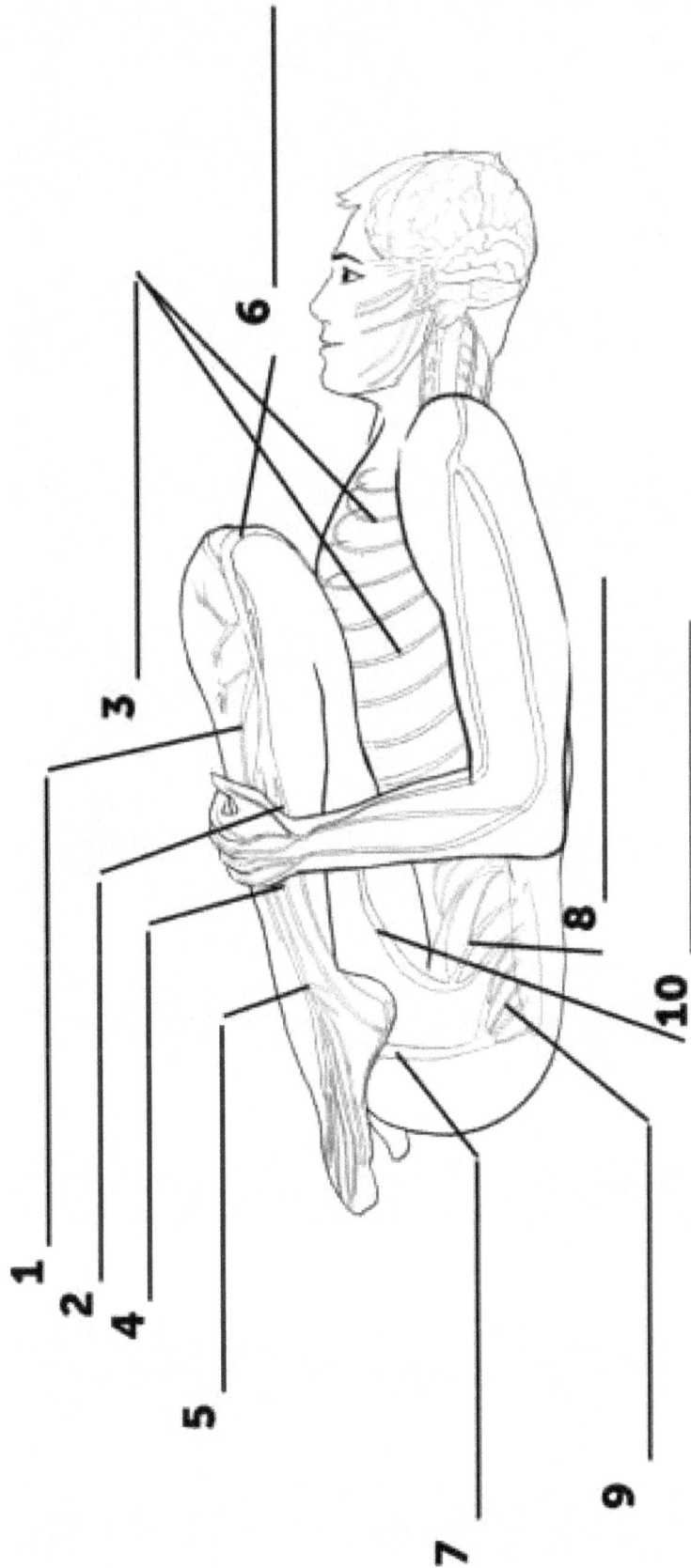

30. WIND REMOVING POSE

1. SAPHENOUS
2. COMMON PERONEAL
3. INTERCOSTALS
4. TIBIAL
5. SUPERFICIAL PERONEAL
6. SCIANTIC
7. SCIANTIC
8. LUMBAR PLEXUS
9. SACRAL PLEXUS
10. FEMORAL

31. RAISED-LEGS POSE

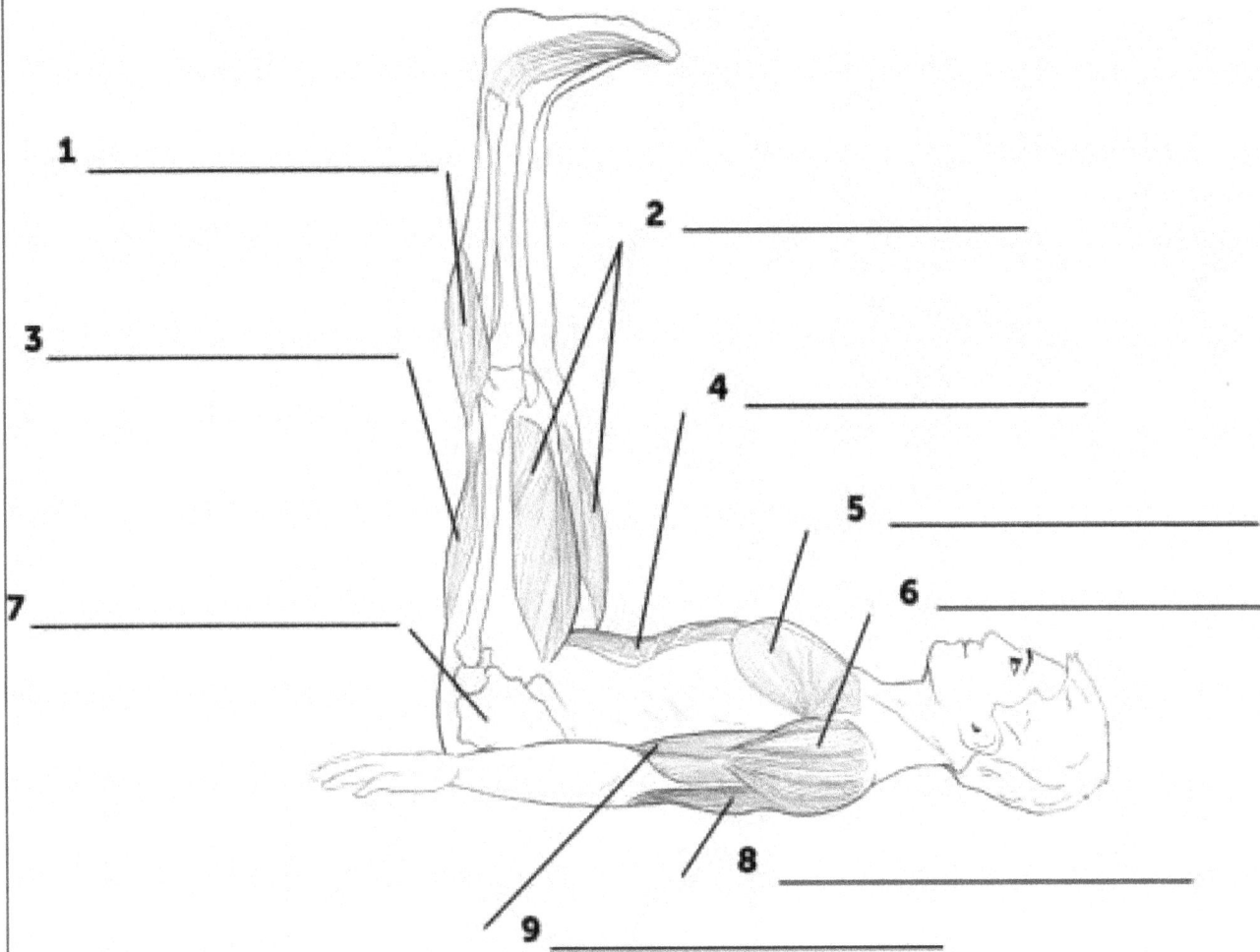

1 _____

2 _____

3 _____

4 _____

5 _____

6 _____

7 _____

8 _____

9 _____

31. RAISED-LEGS POSE

1. GASTROCNEMIUS
2. QUADRICEPS
3. HAMSTRINGS
4. RECTUS ABDOMINIS
5. PECTORALIS MAJOR
6. DELTOID
7. PELVIS
8. TRICEPS BRACHII
9. BICEPS BRACHII

32. CORPSE POSE

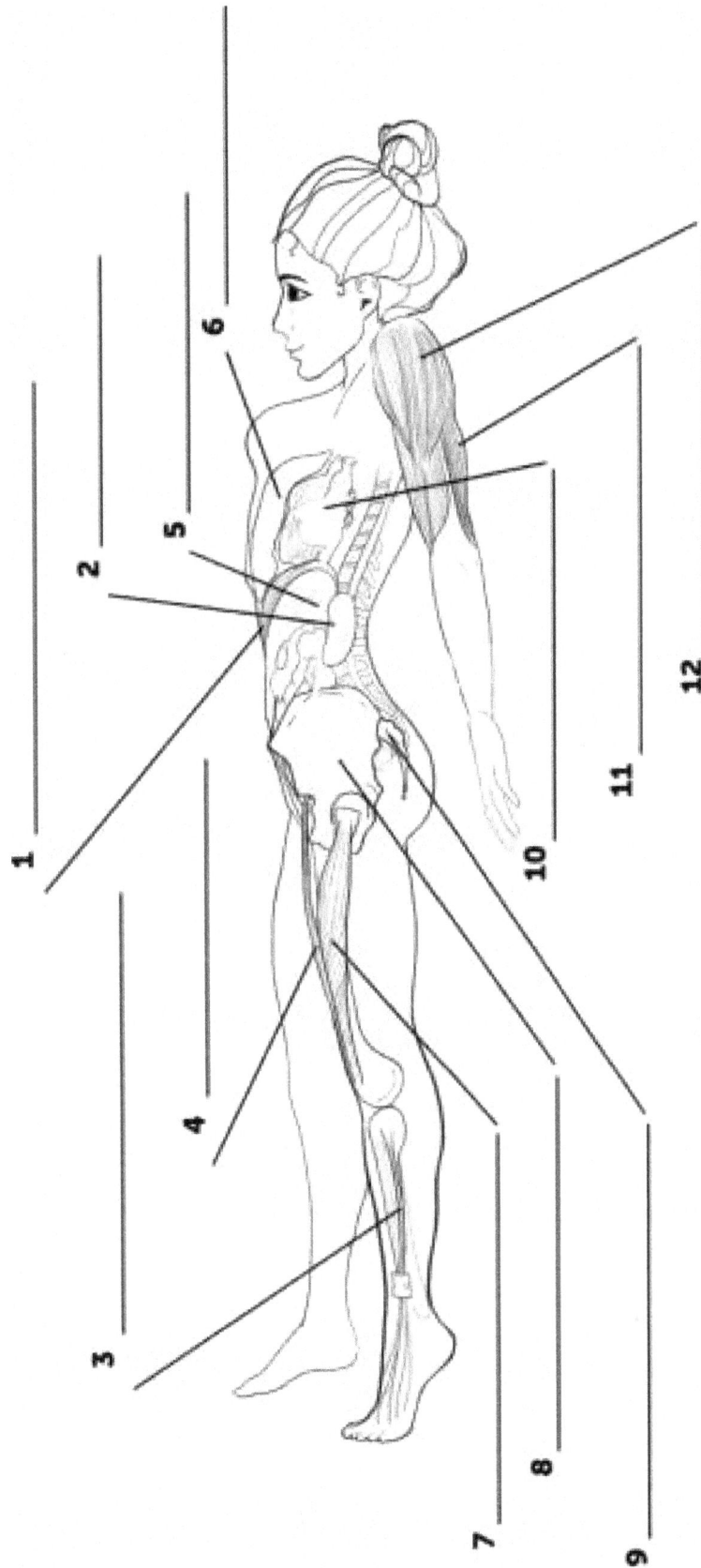

1

2

3

4

5

6

7

8

9

10

11

12

32. CORPSE POSE

1. DIAPHRAGM
2. KIDNEY
3. TIBIALIS ANTERIOR
4. SARTORIUS
5. LIVER
6. LUNGS
7. RECTUS FEMORIS
8. PELVIS
9. SACRUM
10. HEART
11. TRICEPS BRACHII
12. DELTOID

33. RAISED ARMS POSE

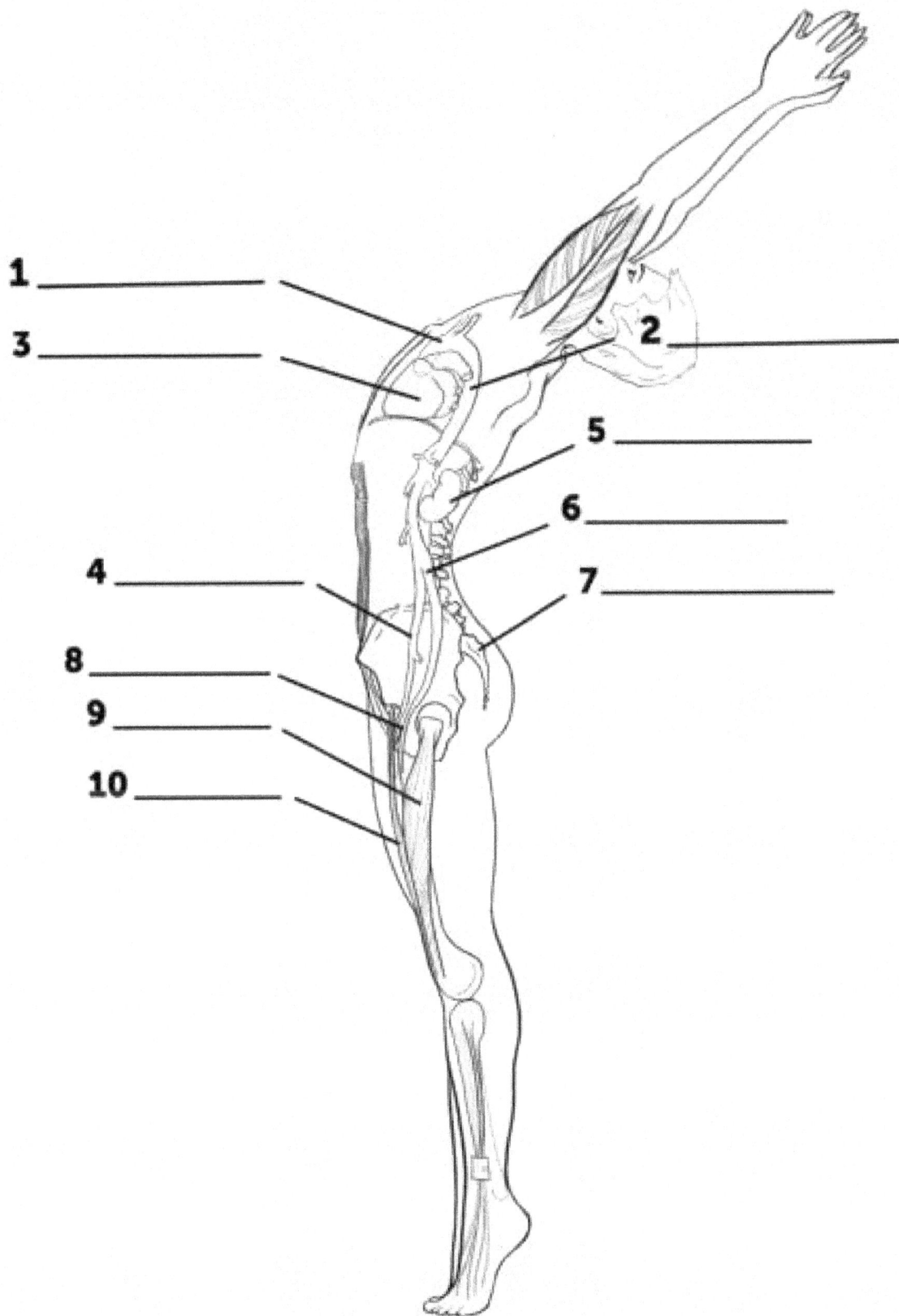

1 _____

3 _____

2 _____

5 _____

6 _____

4 _____

7 _____

8 _____

9 _____

10 _____

33. RAISED ARMS POSE

1. ASCENDING THORACIC AORTA
2. DESCENDING THORACIC AORTA
3. HEART
4. COMMON ILIAC ARTERY
5. KIDNEY
6. ABDOMINAL AORTA
7. SACRUM
8. FEMORAL ARTERY
9. RECTUS FEMORIS
10. SARTORIUS

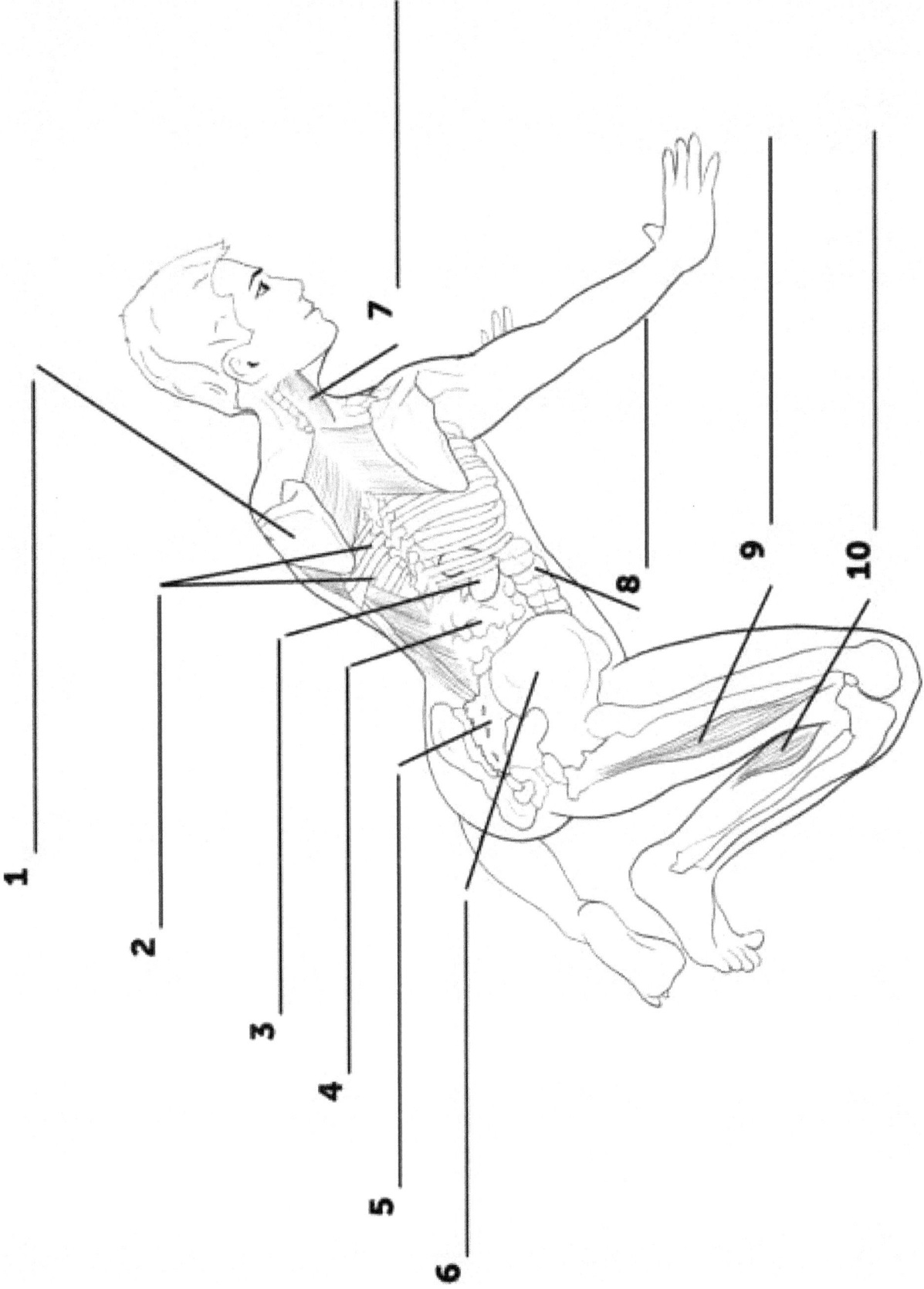

34. FROG POSE

1

2

3

4

5

6

7

8

9

10

34. FROG POSE

1. SCAPULA
2. RIBS
3. KIDNEY
4. COILS OF SMALL INTESTINE
5. SACRUM
6. PELVIS
7. SPLENIUS CAPITIS
8. ASCENDING COLON
9. HAMSTRINGS
10. GASTROCNEMIUS

35. HALF LOTUS POSE

1 _____

2 _____

6 _____

3 _____

4 _____

5 _____

8 _____

7 _____

9 _____

35. HALF LOTUS POSE

1. AORTA
2. HEART
3. LUNGS
4. STOMACH
5. SMALL INTESTINE
6. LIVER
7. LARGE INTESTINE
8. PATELLA
9. GASTROCNEMIUS

36. HAPPY BABY

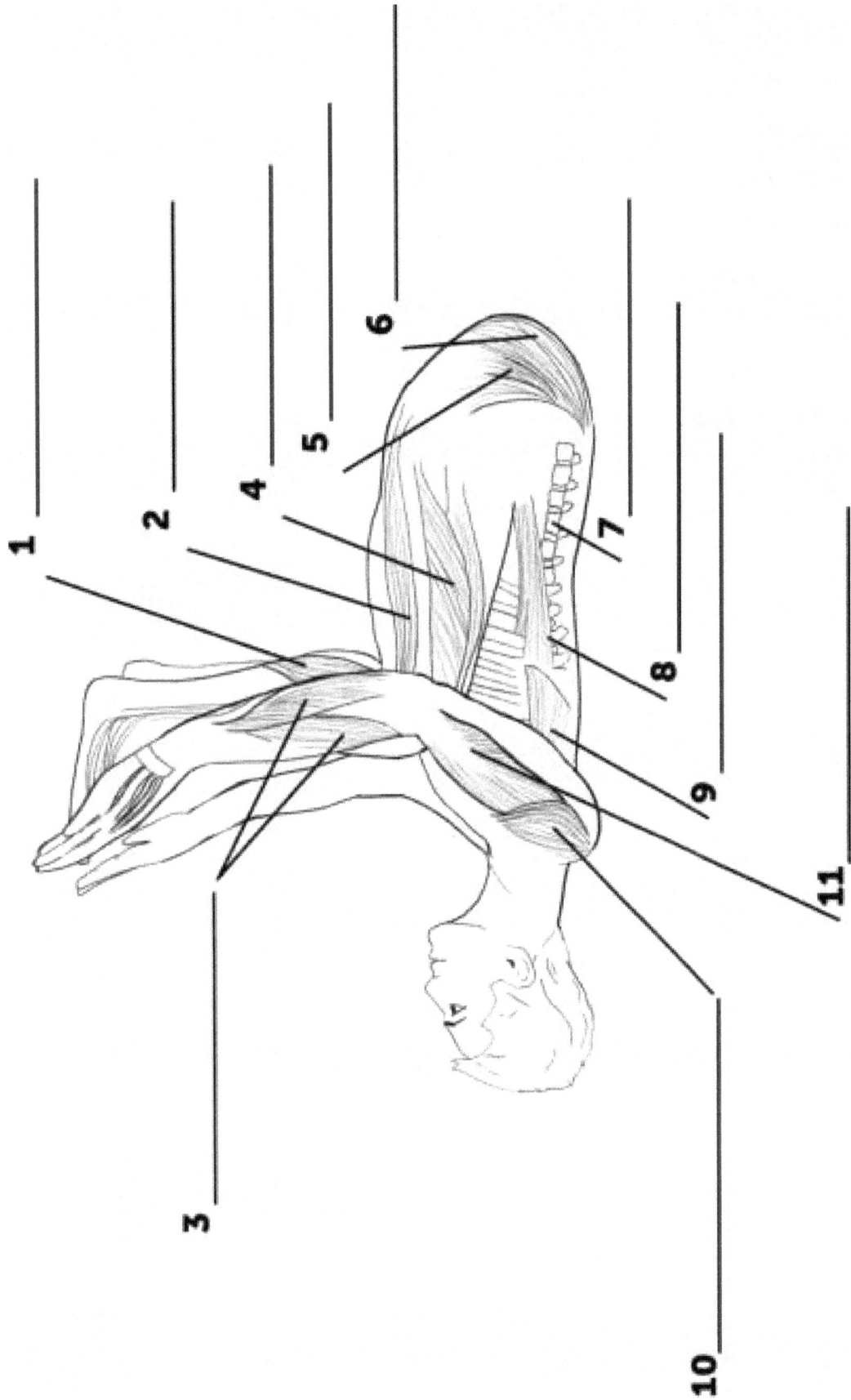

36. HAPPY BABY

1. GASTROCNEMIUS
2. HAMSTRINGS
3. PRONATORS
4. QUADRICEPS
5. PIRIFORMIS
6. GLUTEUS MAXIMUS
7. SPINE
8. ERECTOR SPINAE
9. INFRASPINATUS
10. DELTOID
11. TRICEPS BRACHII

37. LEGS UP THE WALL

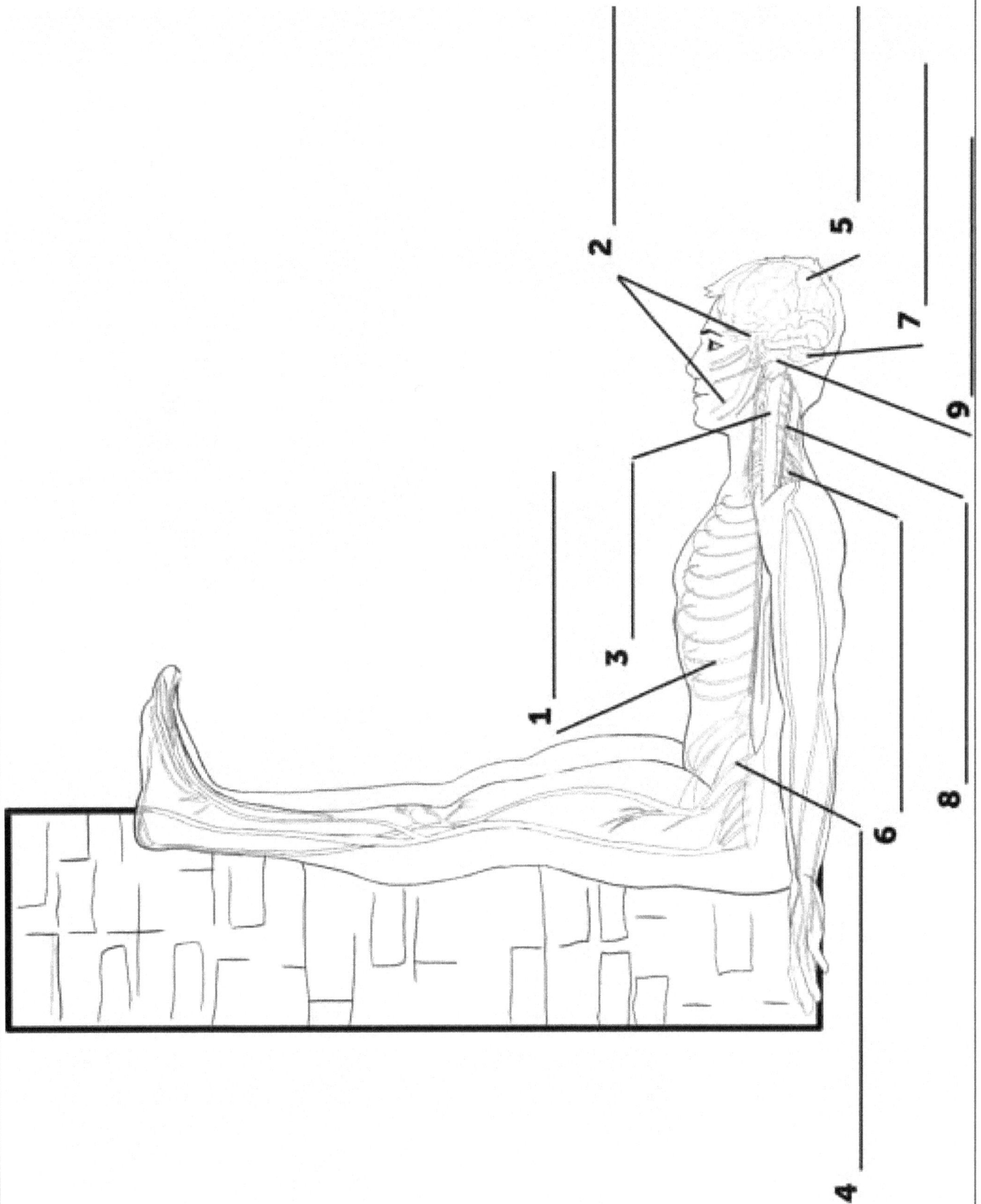

1

2

3

4

5

6

7

8

9

37. LEGS UP THE WALL

1. INTERCOSTALS
2. CRANIAL NERVES
3. SPINAL CORD
4. LUMBAR PLEXUS
5. CEREBRUM
6. BRACHIAL PLEXUS
7. CEREBELLUM
8. VAGUS
9. BRAINSTEM

38. KAPALBHATI POSE

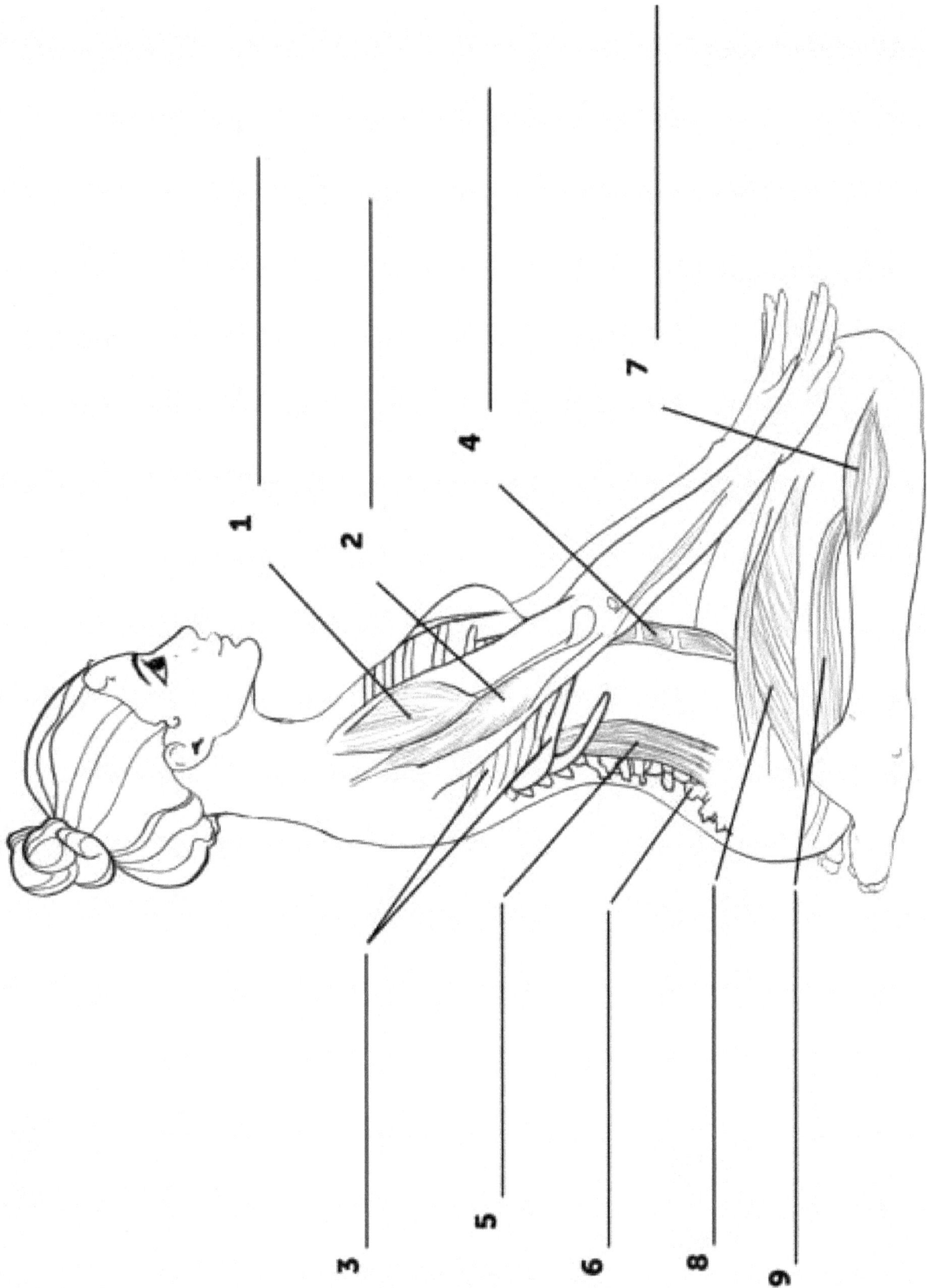

38. KAPALBHATI POSE

1. DELTOID
2. TRICEPS BRACHII
3. RIBS
4. RECTUS ABDOMINIS
5. ERECTOR SPINAE
6. SPINE
7. GASTROCNEMIUS
8. QUADRICEPS
9. HAMSTRINGS

39. LOCUST POSE

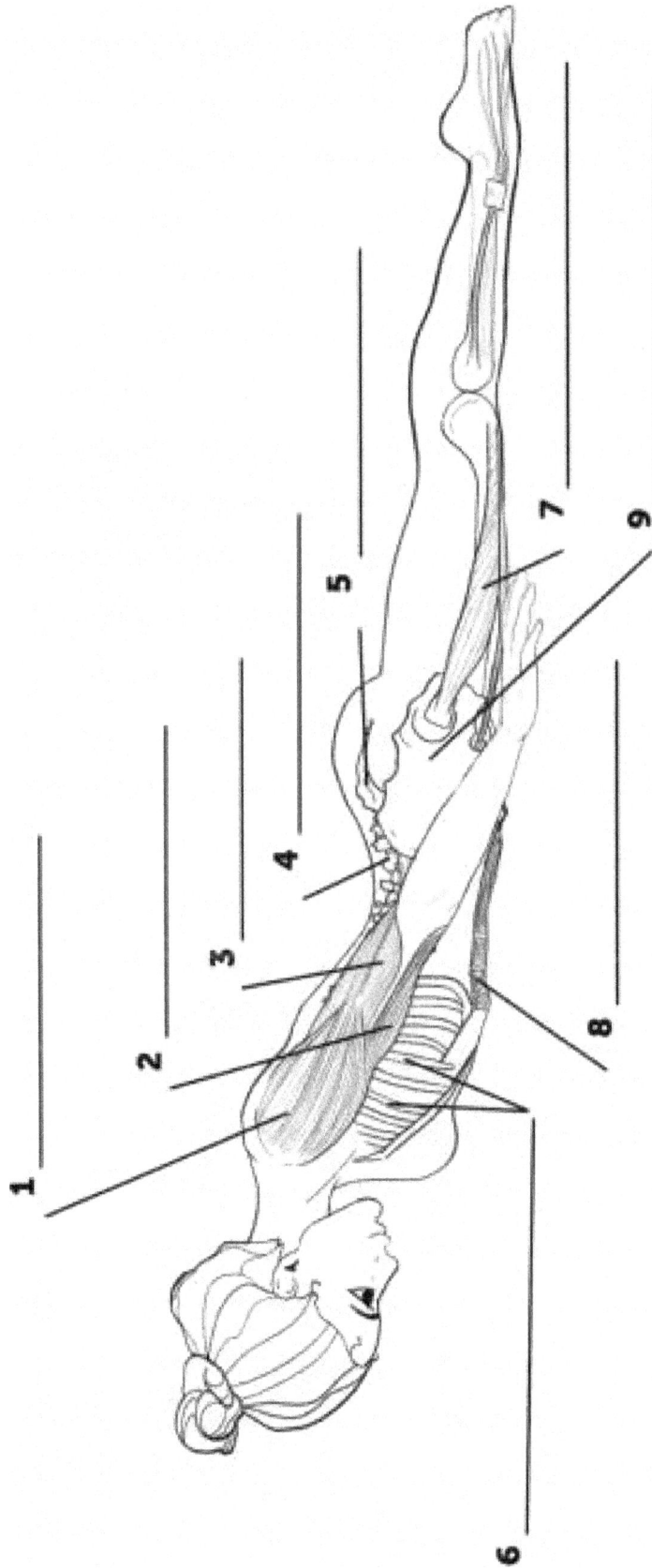

1

2

3

4

5

6

7

8

9

39. LOCUST POSE

1. DELTOID
2. BICEPS BRACHII
3. TRICEPS BRACHII
4. SPINE
5. SACRUM
6. RIBS
7. RECTUS FEMORIS
8. RECTUS ABDOMINIS
9. PELVIS

40. EXTENDED PUPPY POSE

40. EXTENDED PUPPY POSE

1. PIRIFORMIS
2. GLUTEUS MAXIMUS
3. SPINE MUSCLES
4. SPINE
5. HAMSTRINGS
6. RIBS
7. SCAPULA
8. DELTOID
9. GASTROCNEMIUS
10. TRICEPS BRACHII

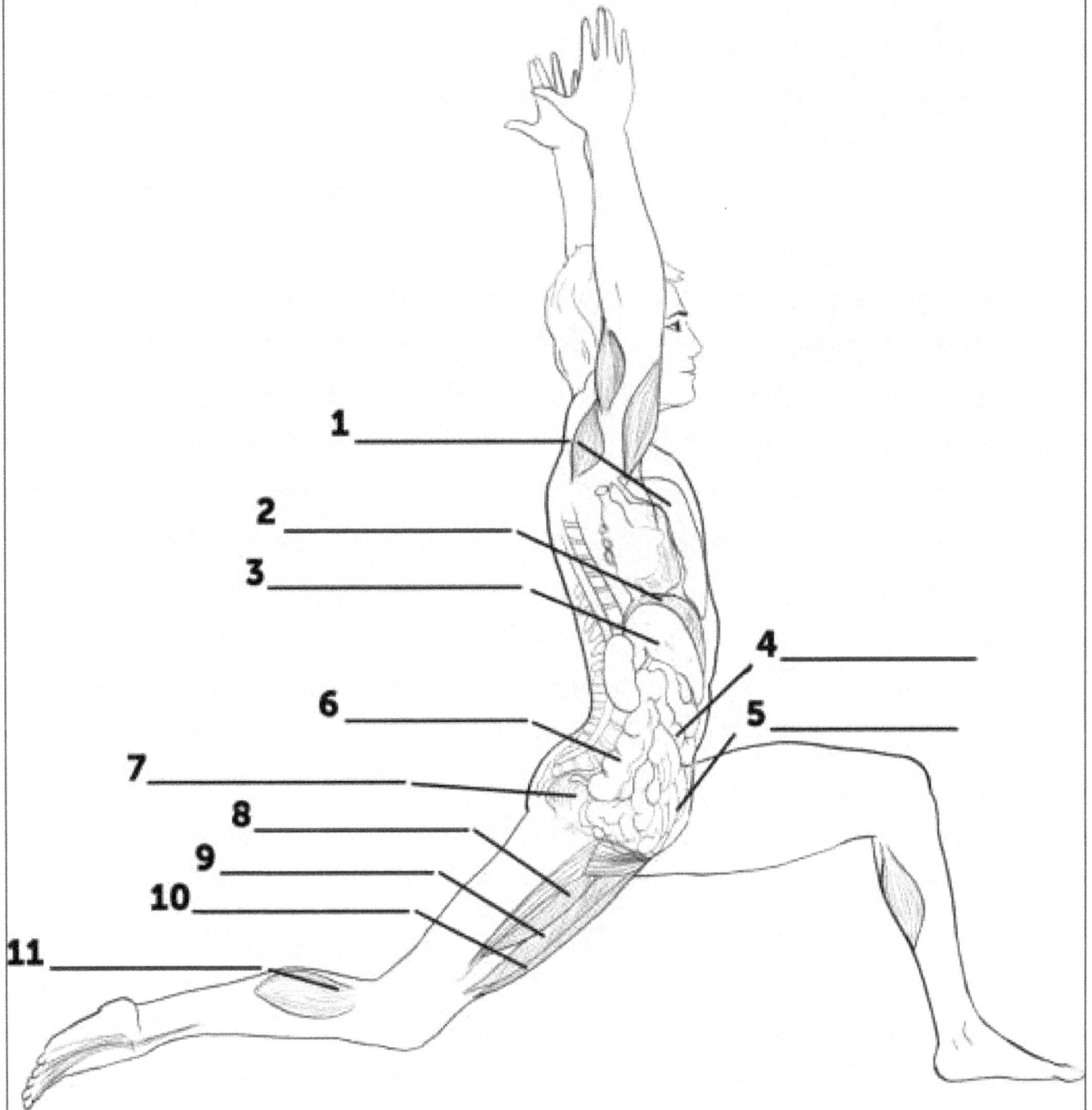

41. LOW LUNGE

1_____

2_____

3_____

4_____

5_____

6_____

7_____

8_____

9_____

10_____

11_____

41. LOW LUNGE

1. LUNGS
2. DIAPHRAGM
3. LIVER
4. TRANSVERSE COLON
5. COILS OF SMALL INTESTINE
6. ASCENDING COLON
7. RECTUM
8. VASTUS LATERALIS
9. RECTUS FEMORIS
10. VASTUS MEDIALIS
11. GASTROCNEMIUS

42. REVOLVED HIGH LUNGE

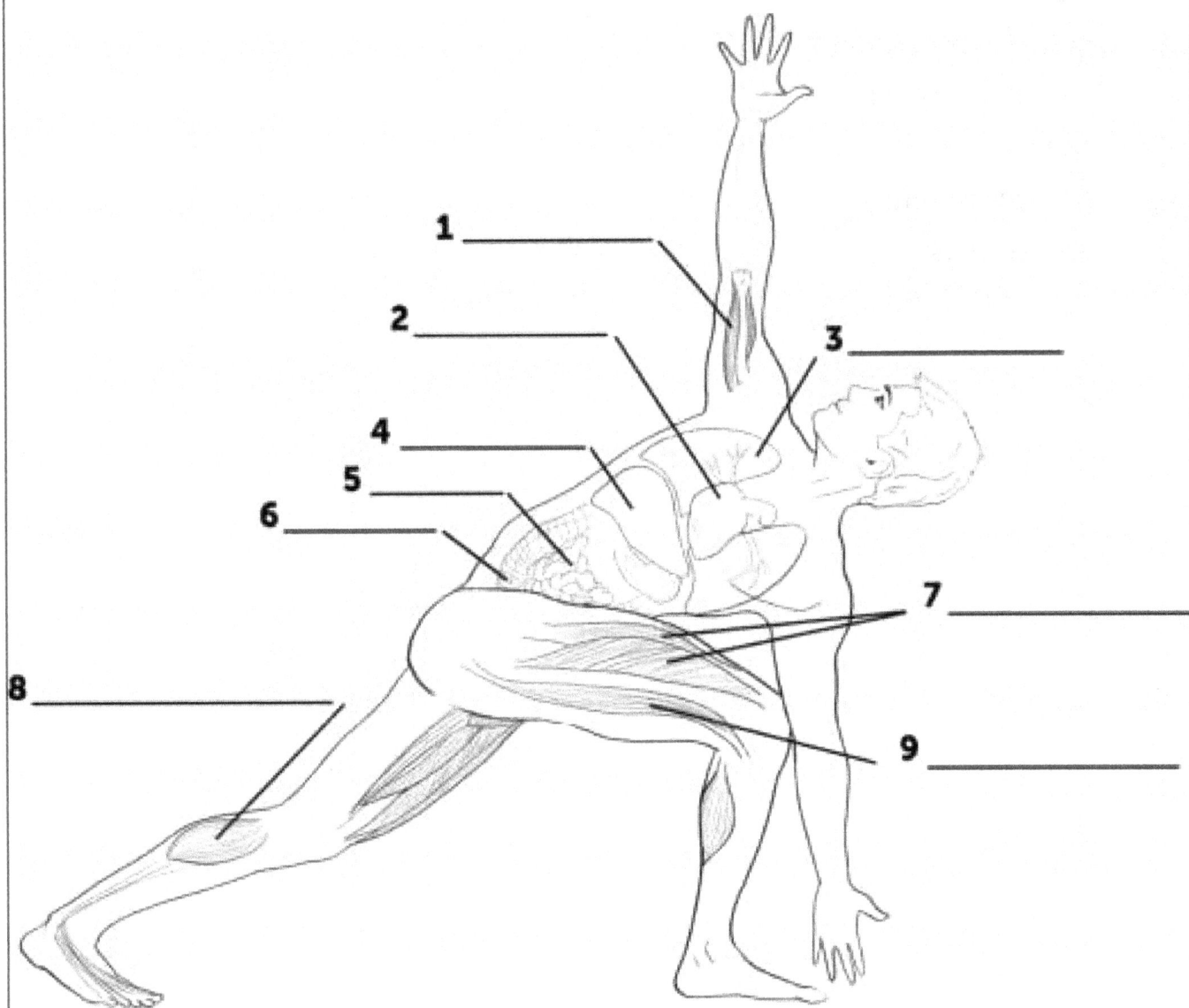

1 _____

2 _____

3 _____

4 _____

5 _____

6 _____

7 _____

8 _____

9 _____

42. REVOLVED HIGH LUNGE

1.   BICEPS BRACHII

2.   HEART

3.   LUNGS

4.   LIVER

5.   COILS OF SMALL INTESTINE

6.   ASCENDING COLON

7.   QUADRICEPS

8.   GASTROCNEMIUS

9.   HAMSTRINGS

43. STANDING WIDE-LEGGED FORWARD FOLD

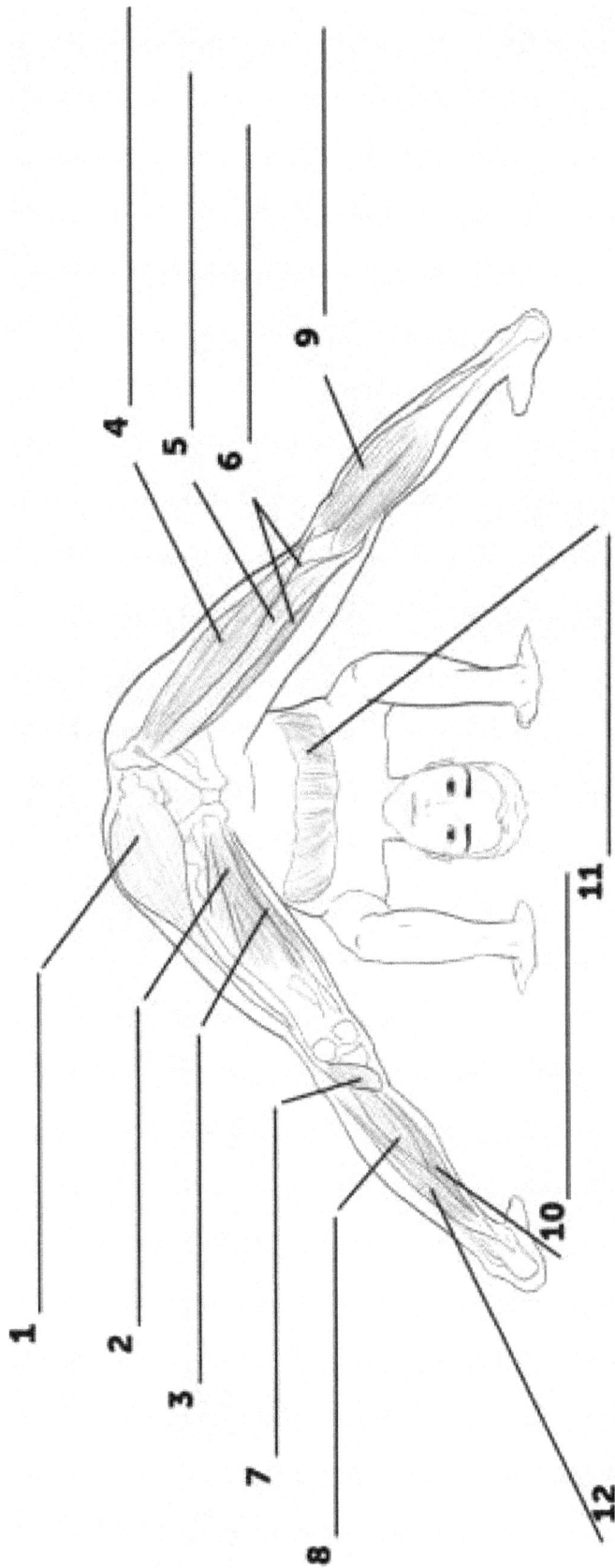

43. STANDING WIDE-LEGGED FORWARD FOLD

1. GLUTEUS MAXIMUS
2. ADDUCTOR MAGNUS
3. GRACILIS
4. BICEPS FEMORIS
5. SEMITENDINOSUS
6. SEMIMEMBRANOSUS
7. POPLITEUS
8. TIBIALIS POSTERIOR
9. GASTROCNEMIUS
10. FLEXOR DIGITORUM LONGUS
11. DIAPHRAGM
12. FLEXOR HALLUCIS LONGUS

44. GODDESS POSE

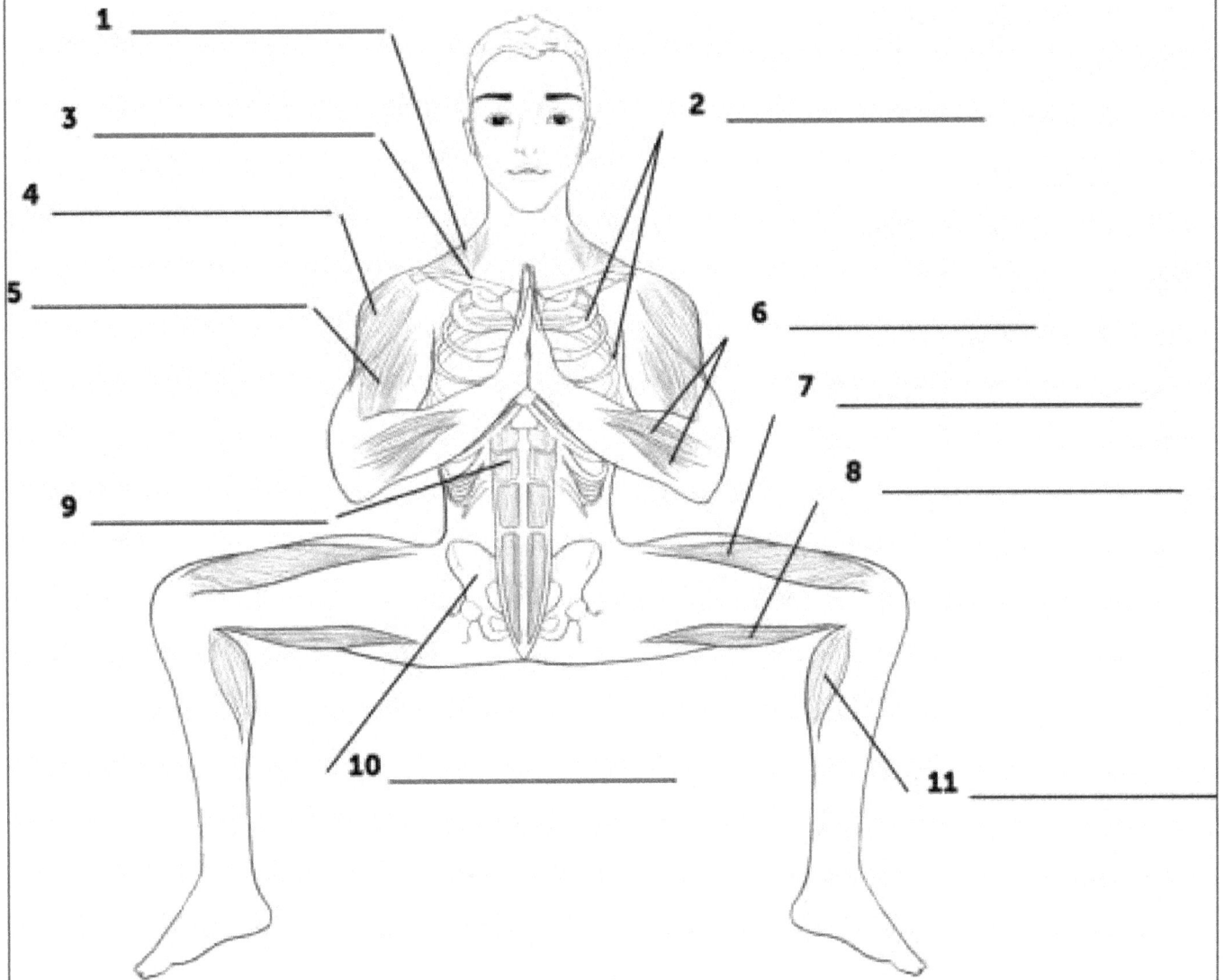

1 _____

3 _____

4 _____

5 _____

2 _____

6 _____

7 _____

8 _____

9 _____

10 _____

11 _____

44. GODDESS POSE

1. TRAPEZIUS
2. RIBS
3. CLAVICLE
4. DELTOID
5. BICEPS BRACHII
6. PRONATORS
7. QUADRICEPS
8. HAMSTRINGS
9. RECTUS ABDOMINIS
10. PELVIS
11. GASTROCNEMIUS

45. BRIDGE ONE LEGGED

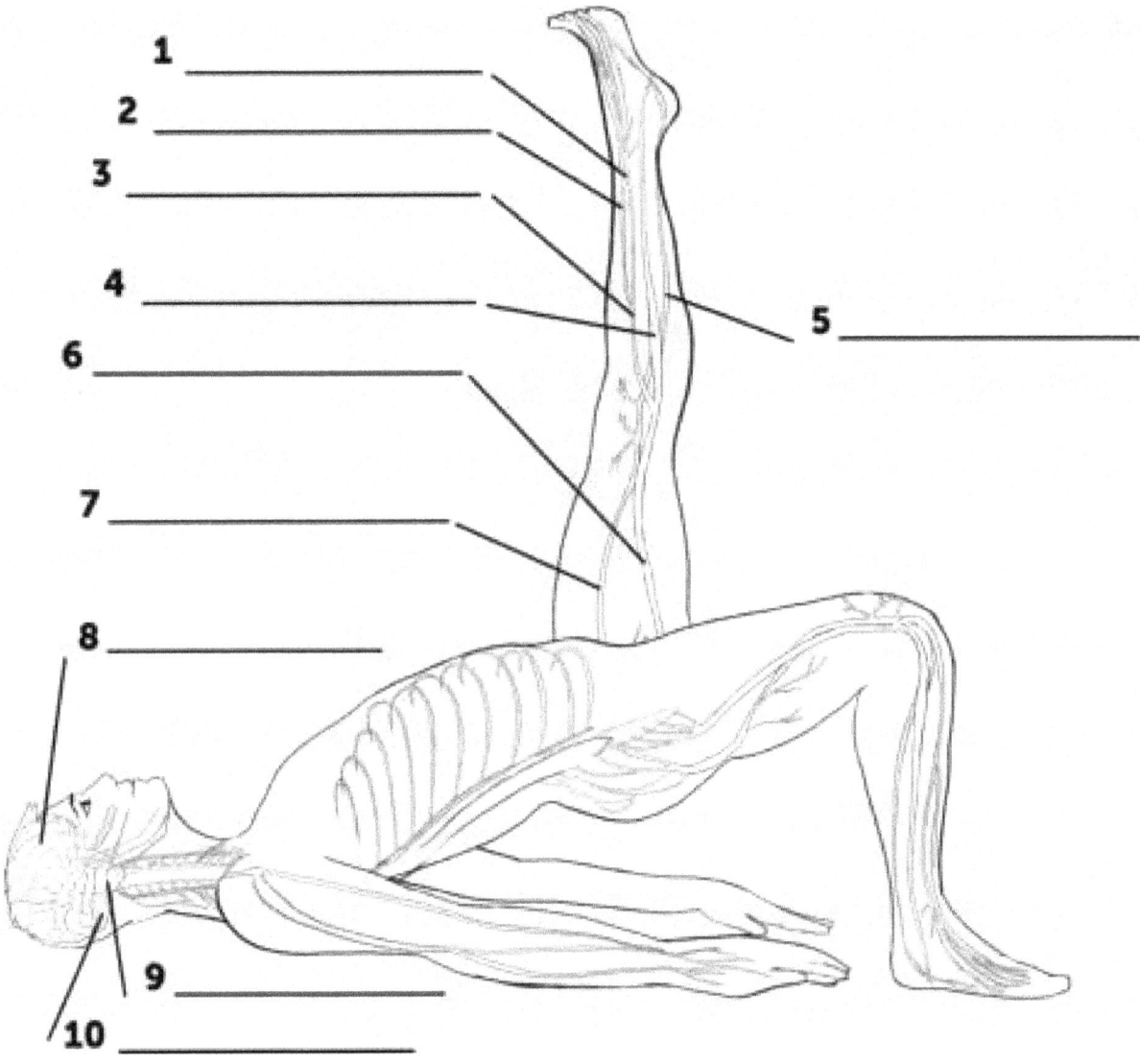

1 _____

2 _____

3 _____

4 _____

5 _____

6 _____

7 _____

8 _____

9 _____

10 _____

45. BRIDGE ONE LEGGED

1. DEEP PERONEAL
2. SUPERFICIAL PERONEAL
3. COMMON PERONEAL
4. TIBIAL
5. SAPHENOUS
6. SCIATIC
7. FEMORAL
8. CEREBRUM
9. BRAINSTEM
10. CEREBELLUM

46. DOUBLE PIGEON POSE

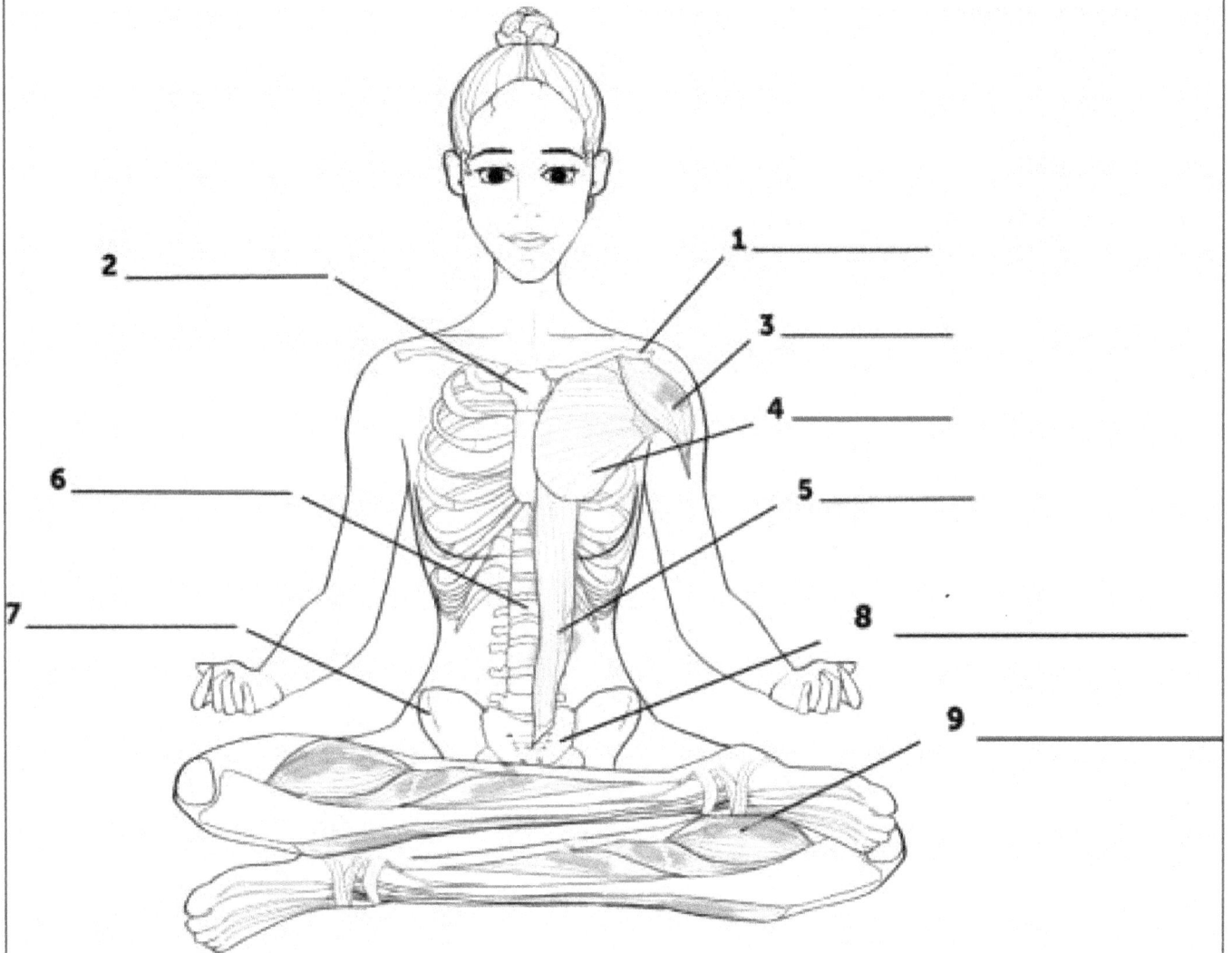

1 _____

2 _____

3 _____

4 _____

5 _____

6 _____

7 _____

8 _____

9 _____

46. DOUBLE PIGEON POSE

1. CLAVICLE
2. STERNUM
3. DELTOID
4. PECTORALIS MAJOR
5. RECTUS ABDOMINIS
6. SPINE
7. PELVIS
8. SACRUM
9. GASTROCNEMIUS

47. SEATED FORWARD BEND

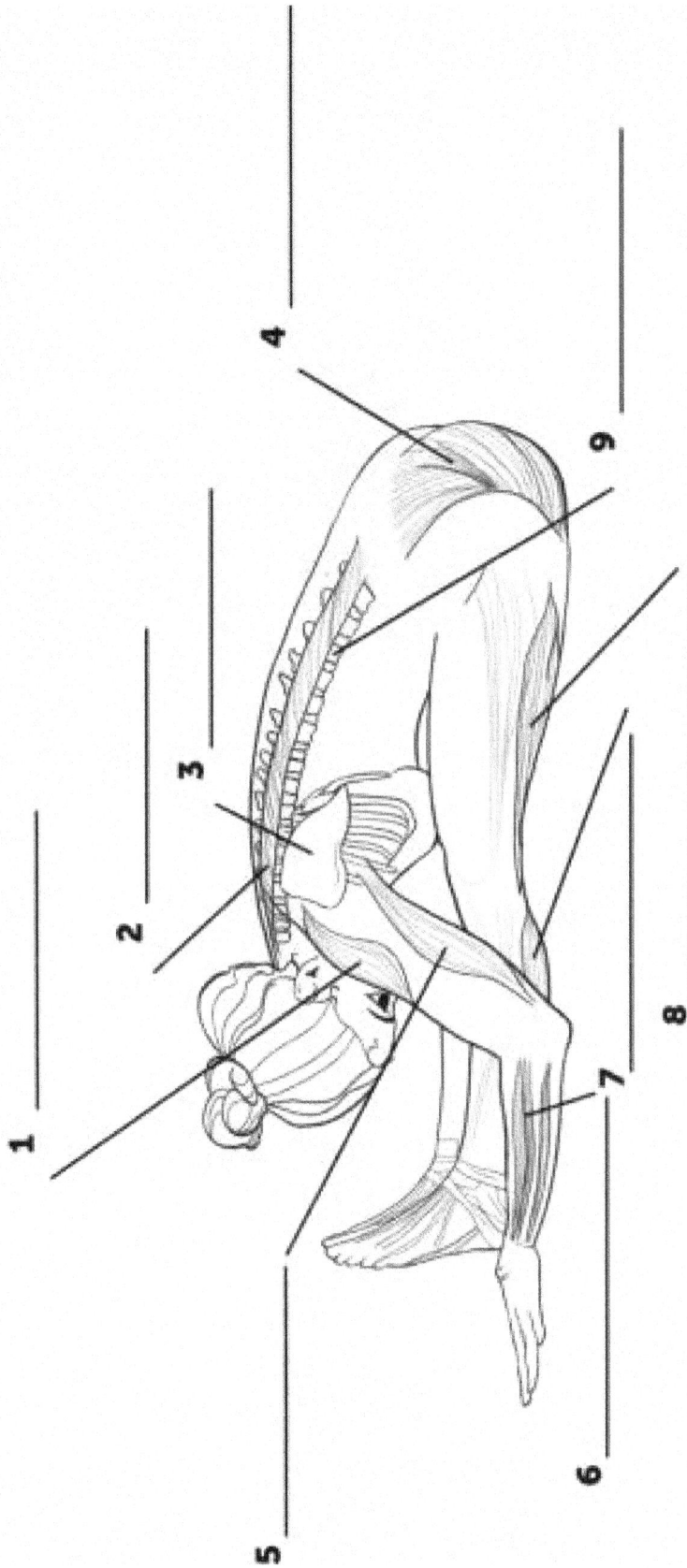

47. SEATED FORWARD BEND

1. DELTOID
2. SPINE MUSCLES
3. SCAPULA
4. PIRIFORMIS
5. TRICEPS BRACHII
6. PRONATORS
7. GASTROCNEMIUS
8. HAMSTRINGS
9. SPINE

48. ONE-LEGGED FORWARD BEND

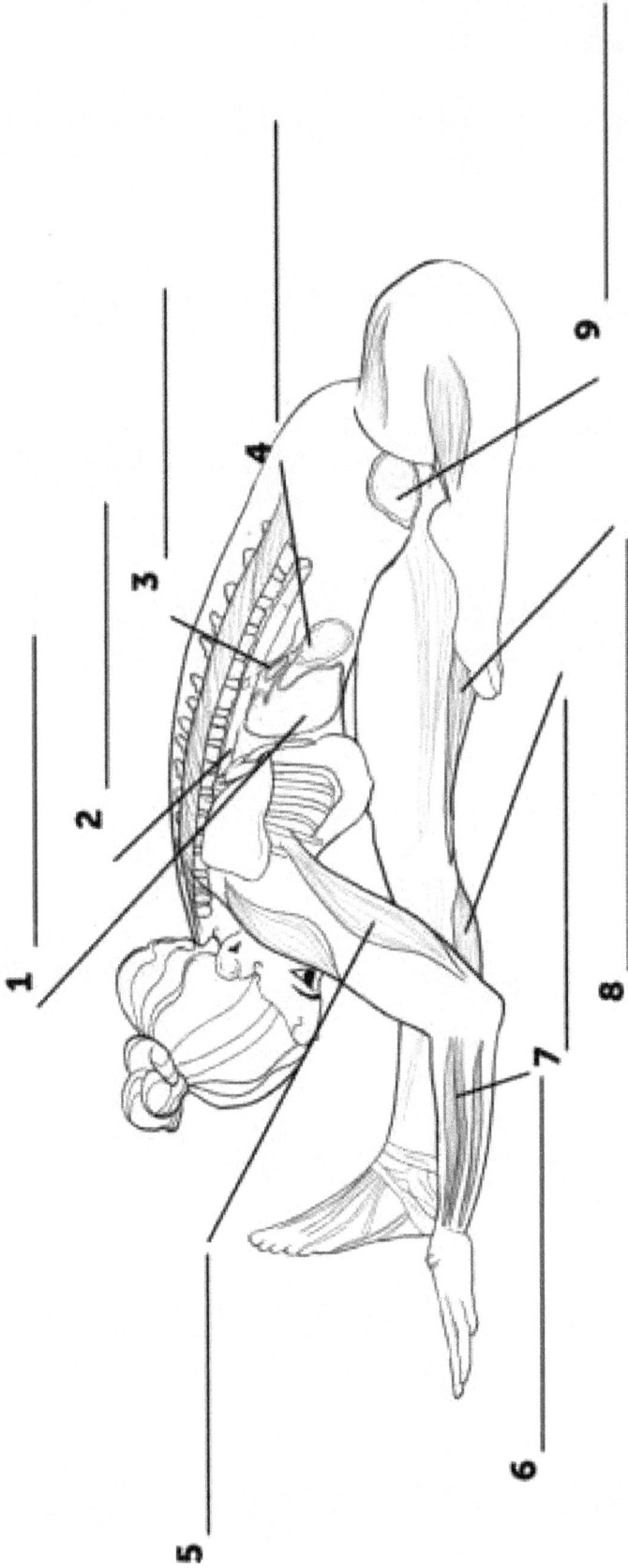

48. ONE-LEGGED FORWARD BEND

1. LIVER
2. ABDOMINAL AORTA
3. PANCREAS
4. STOMACH
5. TRICEPS BRACHII
6. PRONATORS
7. GASTROCNEMIUS
8. HAMSTRINGS
9. URINARY BLADDER

49. KNEES TO CHEST

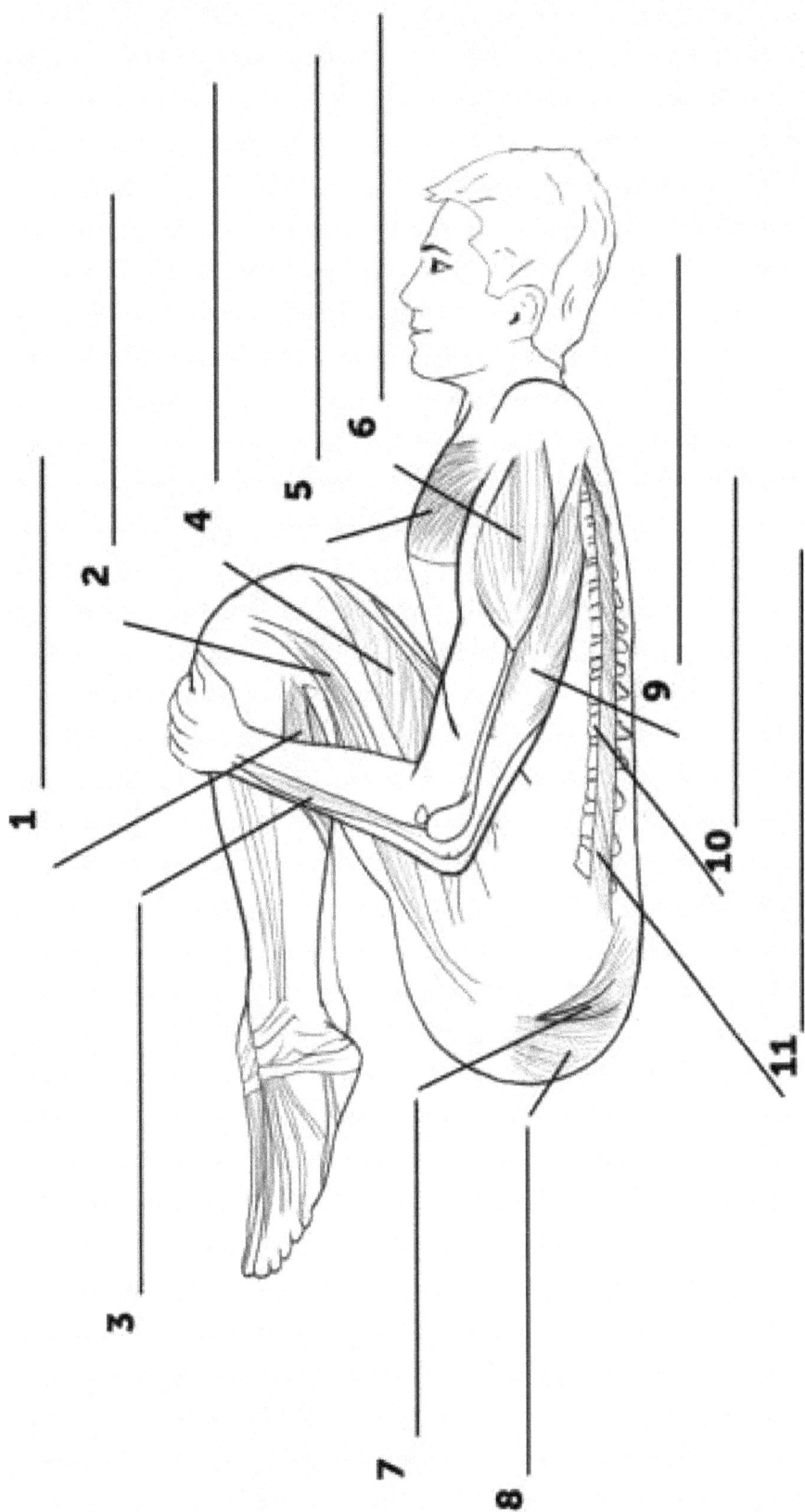

1

2

3

4

5

6

7

8

9

10

11

49. KNEES TO CHEST

1. GASTROCNEMIUS
2. HAMSTRINGS
3. PRONATORS
4. QUADRICEPS
5. PECTORALIS MAJOR
6. DELTOID
7. PIRIFORMIS
8. GLUTEUS MAXIMUS
9. TRICEPS BRACHII
10. SPINE
11. SPINE MUSCLES

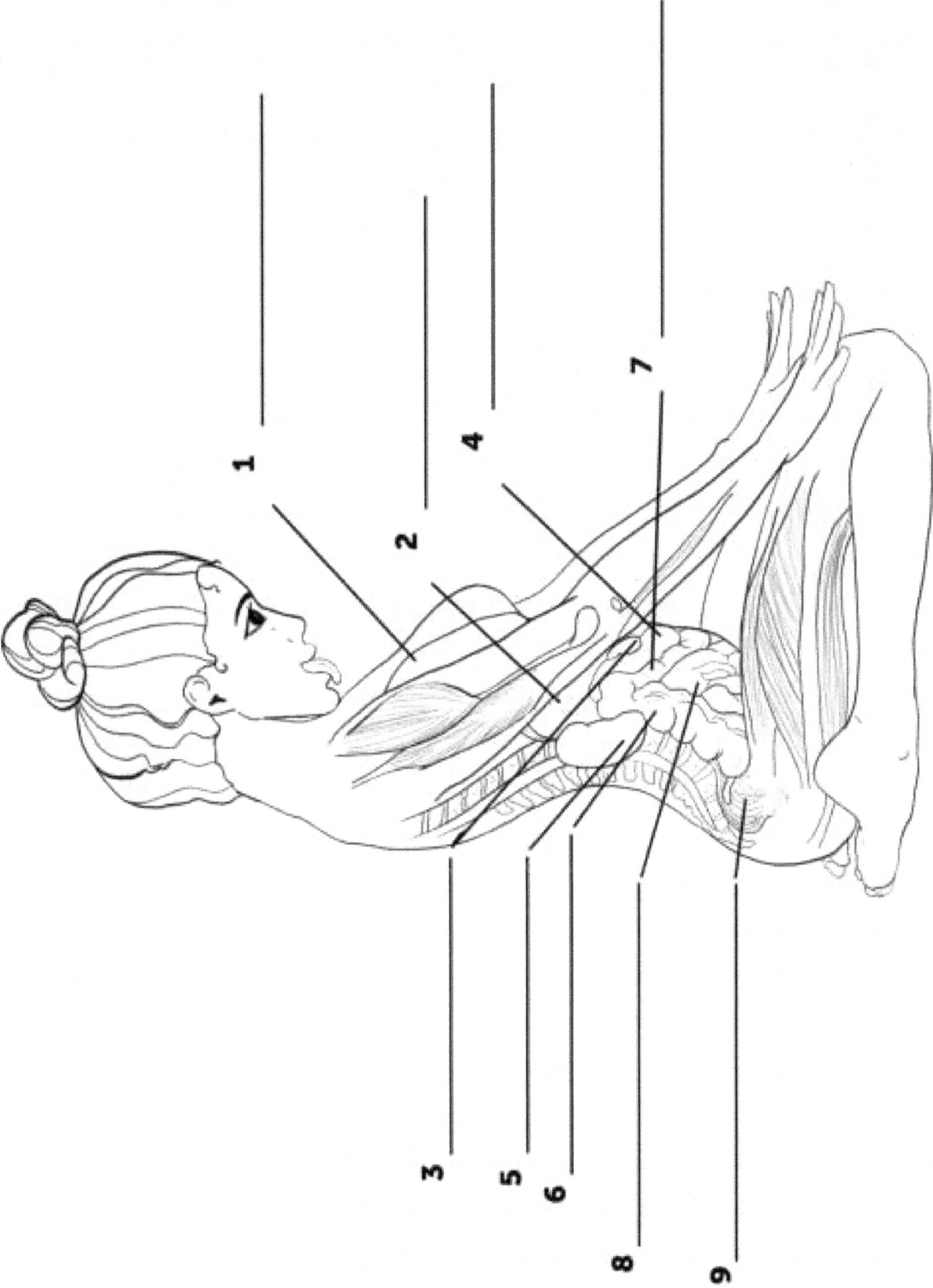

50. LION POSE

1

2

3

4

5

6

7

8

9

50. LION POSE

1. LUNGS
2. LIVER
3. GALLBLADDER
4. STOMACH
5. KIDNEY
6. ASCENDING COLON
7. TRANSVERSE COLON
8. COILS OF SMALL INTESTINE
9. RECTUM

51. HALF KNEES TO CHEST

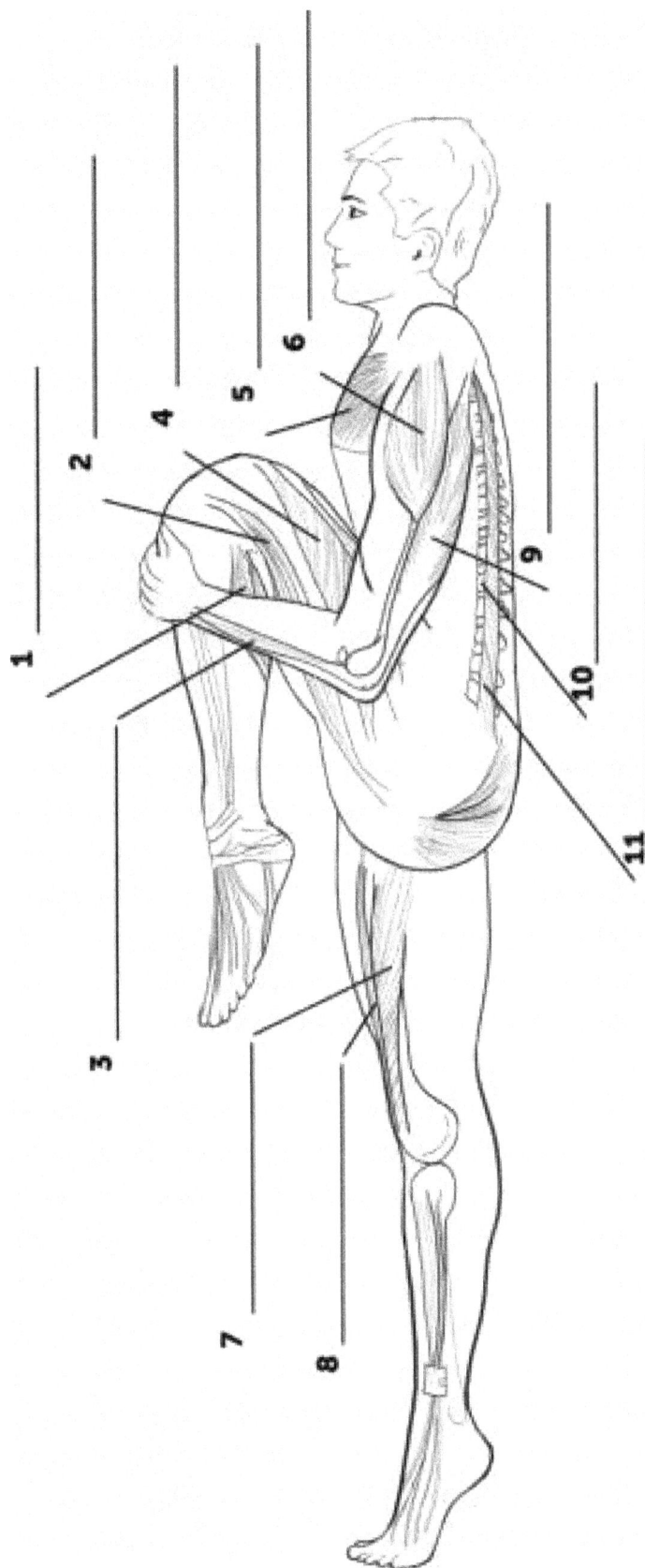

51. HALF KNEES TO CHEST

1. GASTROCNEMIUS
2. HAMSTRINGS
3. PRONATORS
4. QUADRICEPS
5. PECTORALIS MAJOR
6. DELTOID
7. RECTUS FEMORIS
8. SARTORIUS
9. TRICEPS BRACHII
10. SPINE
11. SPINE MUSCLES

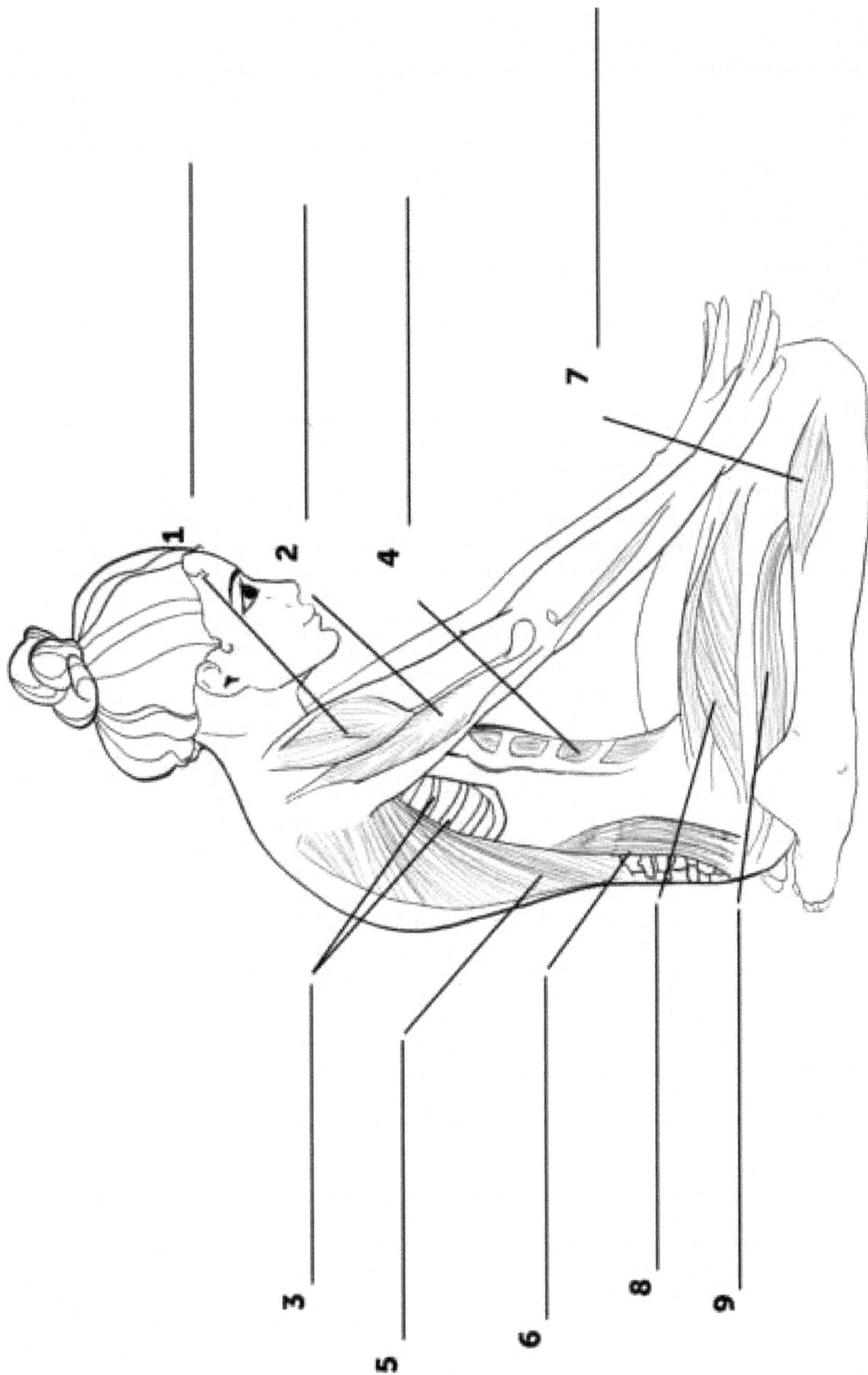

52. CAT SEATED

1

2

4

7

3

5

6

8

9

52. CAT SEATED

1. DELTOID
2. TRICEPS BRACHII
3. RIBS
4. RECTUS ABDOMINIS
5. LATISSIMUS DORSI
6. ERECTOR SPINAE
7. GASTROCNEMIUS
8. QUADRICEPS
9. HAMSTRINGS

53. STANDING KNEES TO CHEST

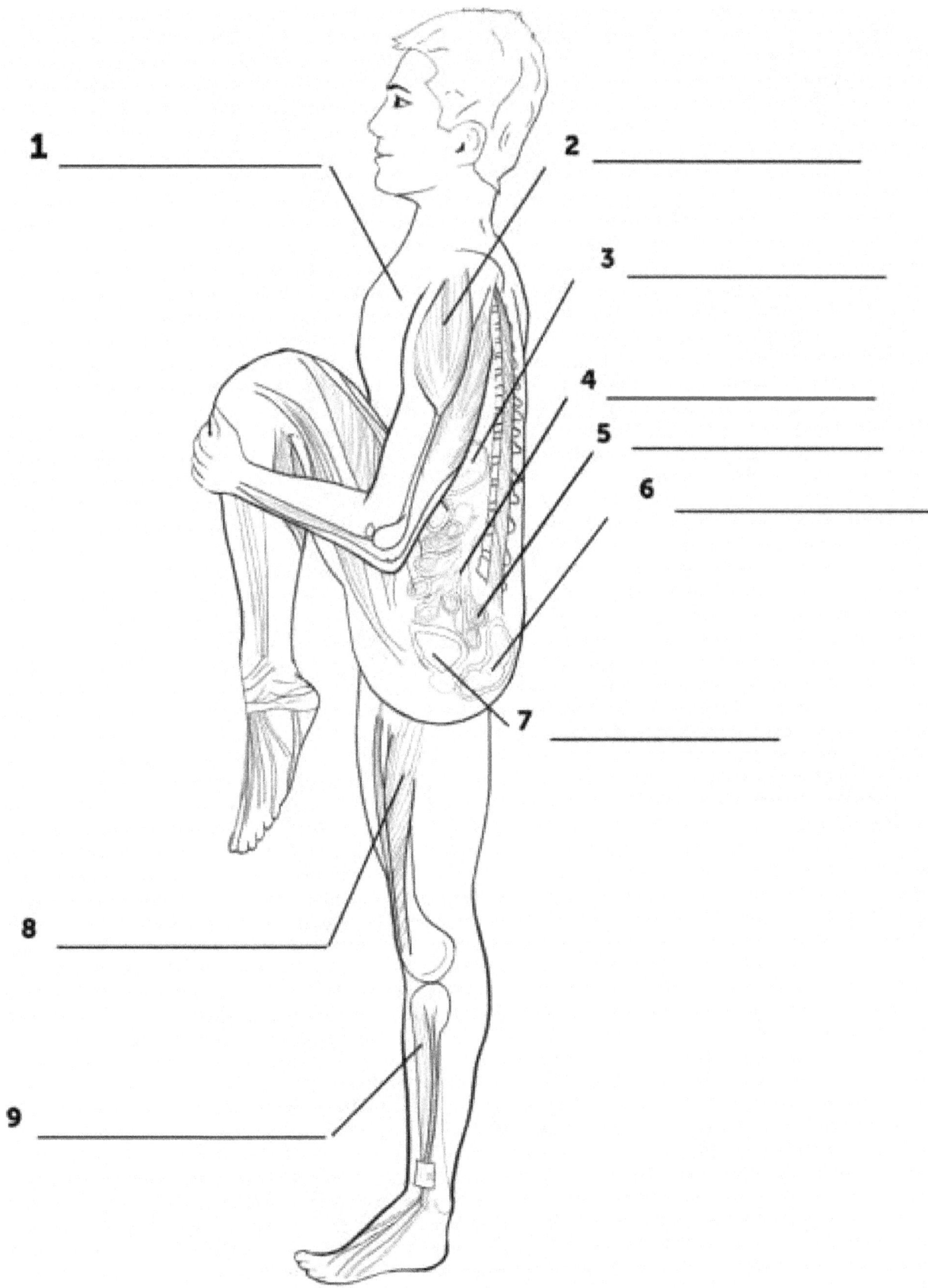

1 _____

2 _____

3 _____

4 _____

5 _____

6 _____

7 _____

8 _____

9 _____

53. STANDING KNEES TO CHEST

1. CHEST
2. DELTOID
3. STOMACH
4. MESENTERY OF THE SMALL INTESTINE
5. COILS OF SMALL INTESTINE
6. RECTUM
7. URINARY BLADDER
8. RECTUS FEMORIS
9. TIBIALIS ANTERIOR

54. STANDING HALF LOTUS POSE

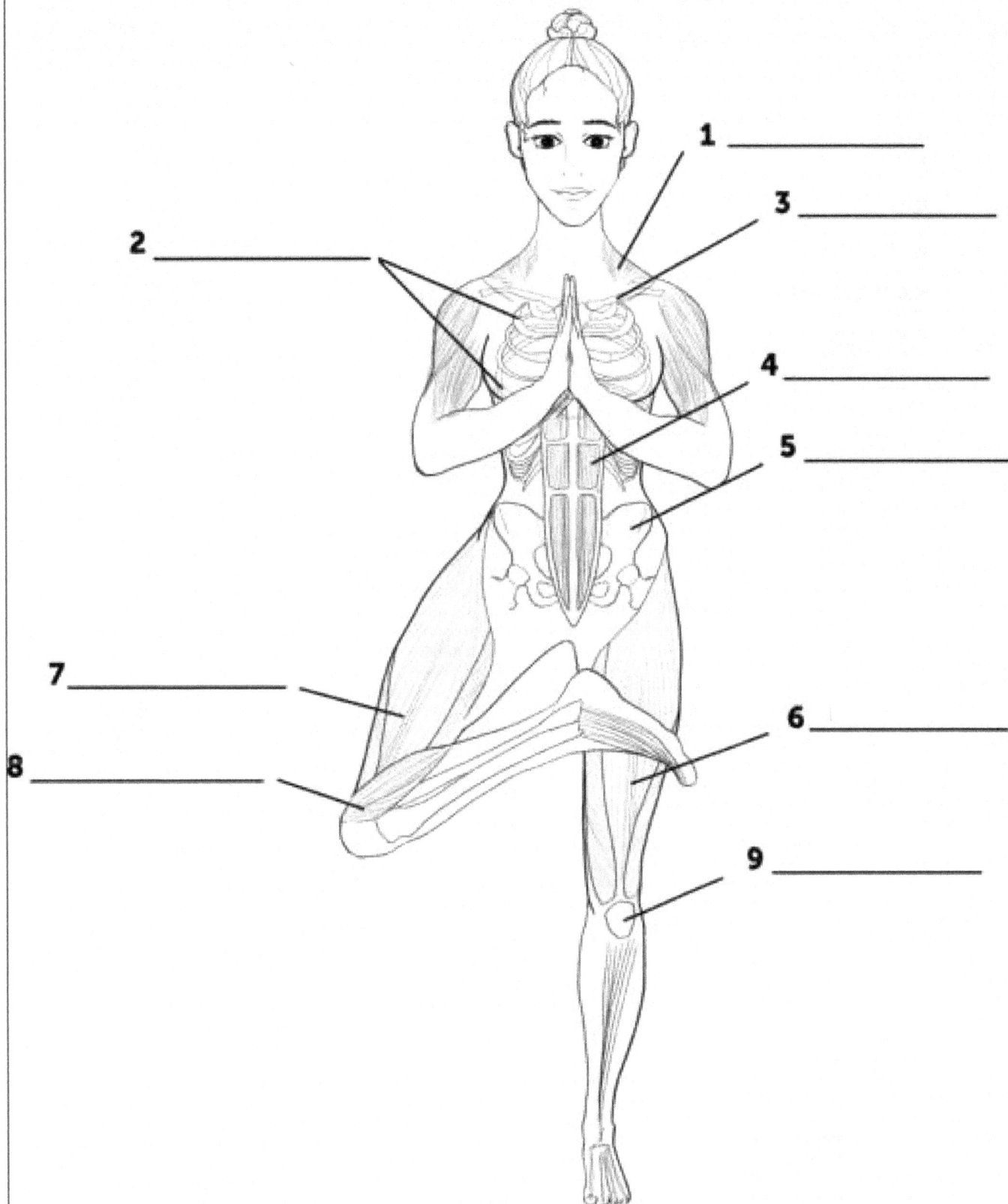

1 _____

3 _____

2 _____

4 _____

5 _____

7 _____

6 _____

8 _____

9 _____

54. STANDING HALF LOTUS POSE

1. TRAPEZIUS
2. RIBS
3. CLAVICLE
4. RECTUS ABDOMINIS
5. PELVIS
6. QUADRICEPS
7. HAMSTRINGS
8. GASTROCNEMIUS
9. PATELLA

YOGA POSES FOR INTERMEDIATES

55. SIDE PLANK POSE

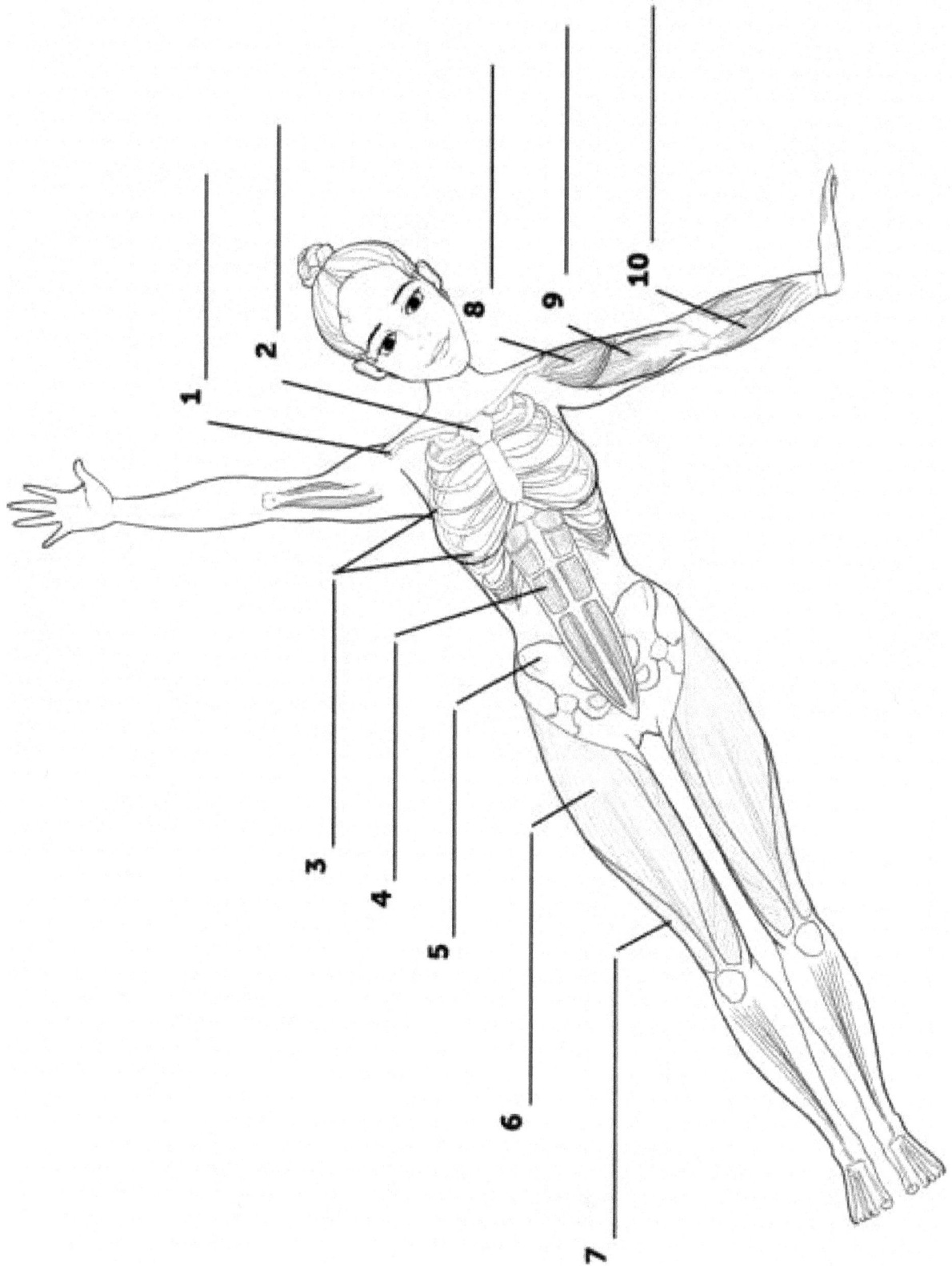

1

2

3

4

5

6

7

8

9

10

55. SIDE PLANK POSE

1. COLLARBONE
2. STERNUM
3. RIBS
4. RECTUS ABDOMINIS
5. PELVIS
6. QUADRICEPS
7. VASTUS LATERALIS
8. DELTOID
9. BICEPS BRACHII
10. PRONATORS

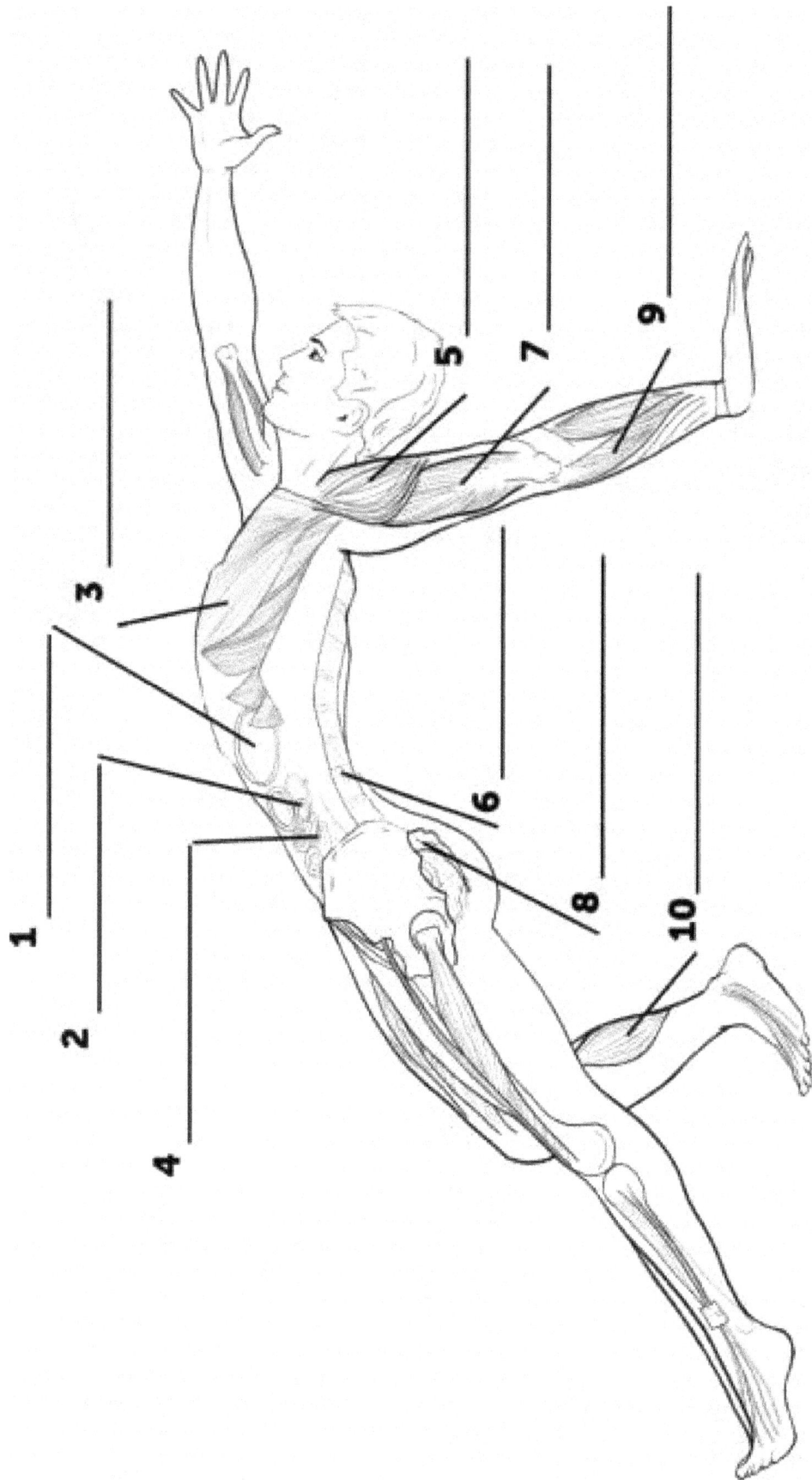

56. WILD THING

1

2

3

4

5

6

7

8

9

10

56. WILD THING

1. STOMACH

2. COILS OF SMALL INTESTINE

3. PECTORALIS MAJOR

4. MESENTERY OF SMALL INTESTINE

5. DELTOID

6. SPINE

7. BICEPS BRACHII

8. SACRUM

9. PRONATORS

10. GASTROCNEMIUS

57. HALF FROG POSE

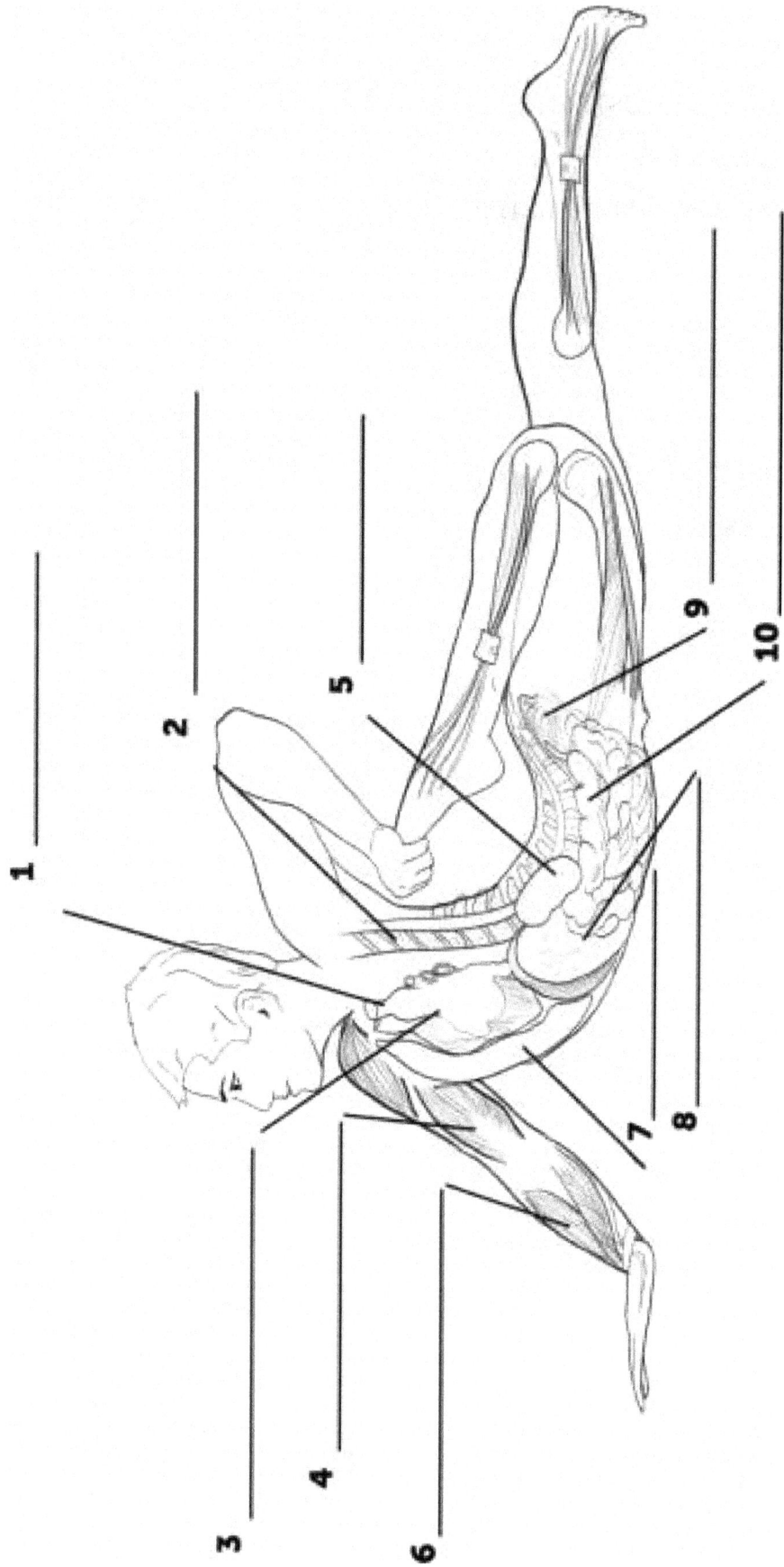

57. HALF FROG POSE

1. AORTA
2. SPINE
3. HEART
4. BICEPS BRACHII
5. KIDNEY
6. PRONATORS
7. LUNGS
8. LIVER
9. RECTUM
10. ASCENDING COLON

58. COMPASS POSE

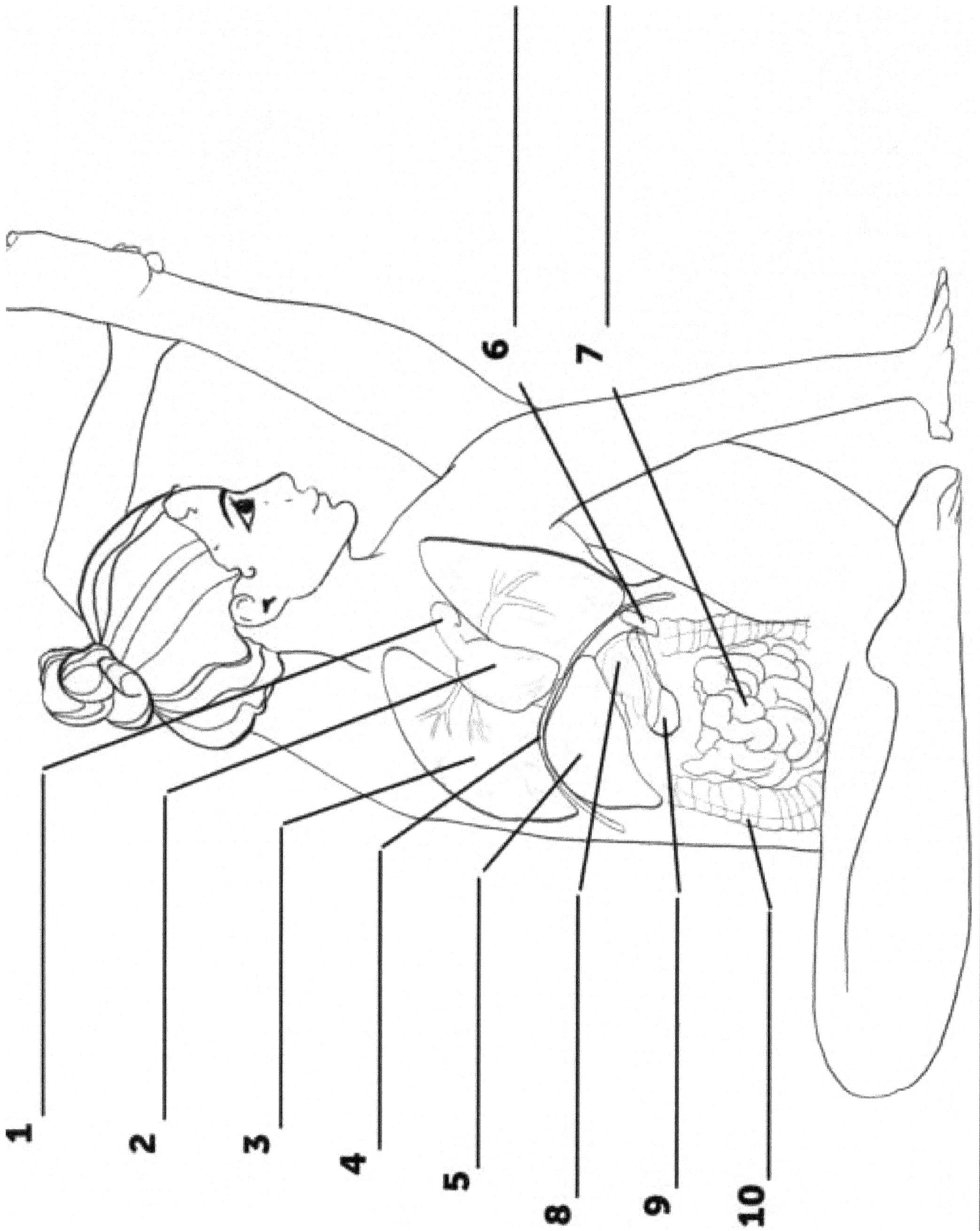

1

2

3

4

5

6

7

8

9

10

58. COMPASS POSE

1. AORTA
2. HEART
3. LUNGS
4. DIAPHRAGM
5. LIVER
6. GALLBLADDER
7. COILS OF SMALL INTESTINE
8. STOMACH
9. PANCREAS
10. ASCENDING COLON

59. MARICHI'S POSE I

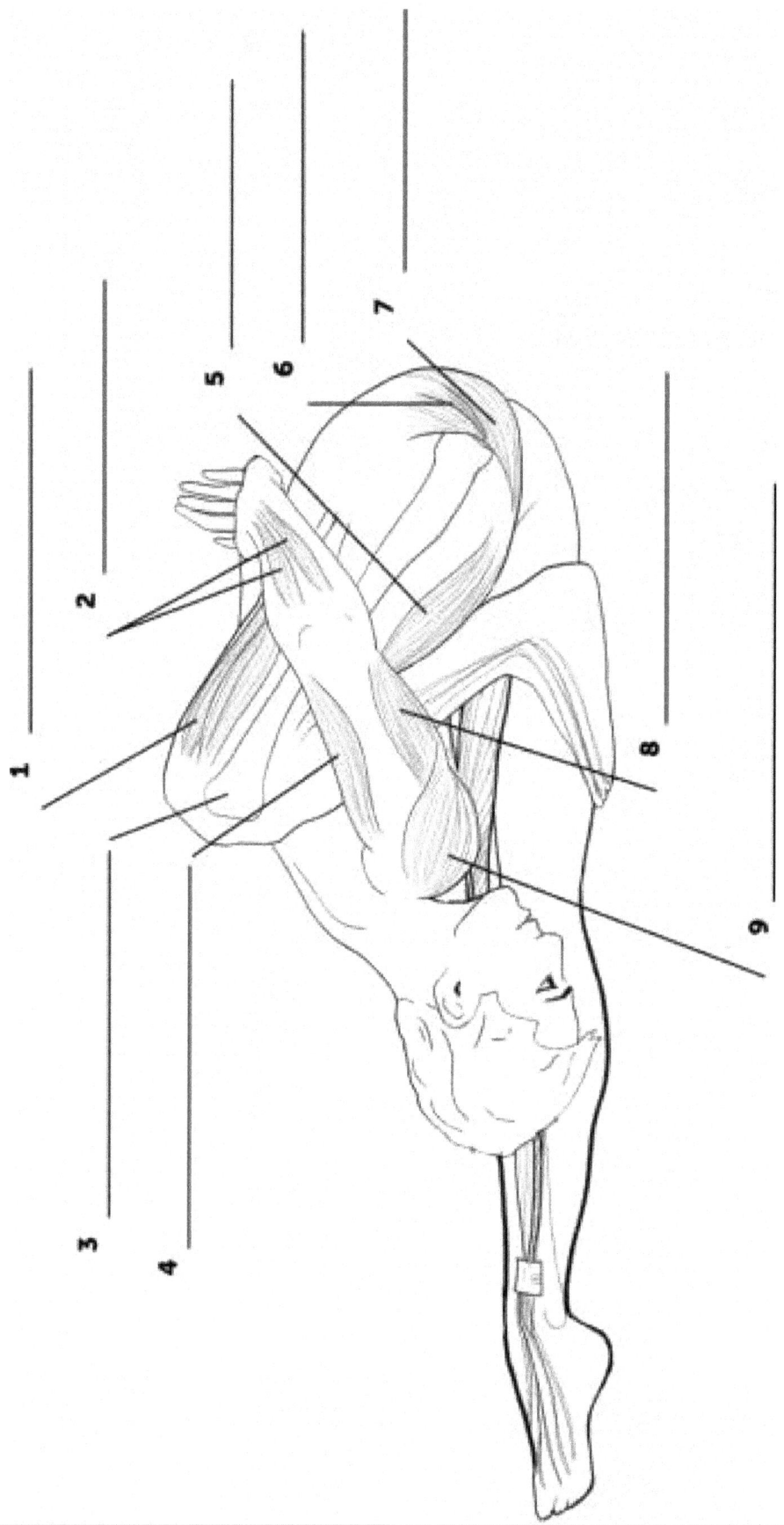

59. MARICHI'S POSE I

1.   QUADRICEPS

2.   PRONATORS

3.   FEMUR

4.   BICEPS BRACHII

5.   HAMSTRINGS

6.   PIRIFORMIS

7.   GLUTEUS MAXIMUS

8.   TRICEPS BRACHII

9.   DELTOID

60. MARICHI'S POSE II

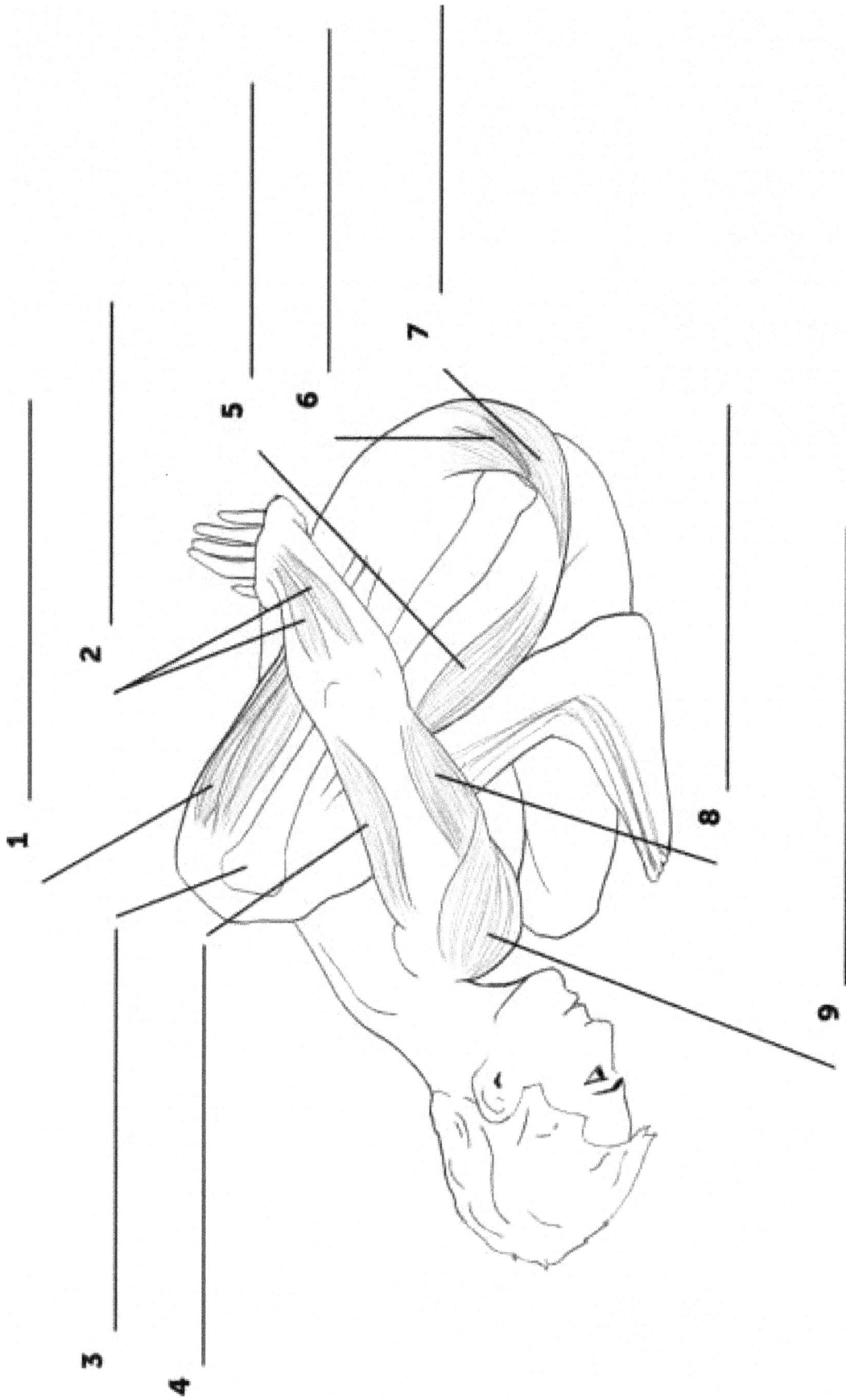

60. MARICHI'S POSE II

1. QUADRICEPS
2. PRONATORS
3. FEMUR
4. BICEPS BRACHII
5. HAMSTRINGS
6. PIRIFORMIS
7. GLUTEUS MAXIMUS
8. TRICEPS BRACHII
9. DELTOID

61. MARICHI'S POSE III

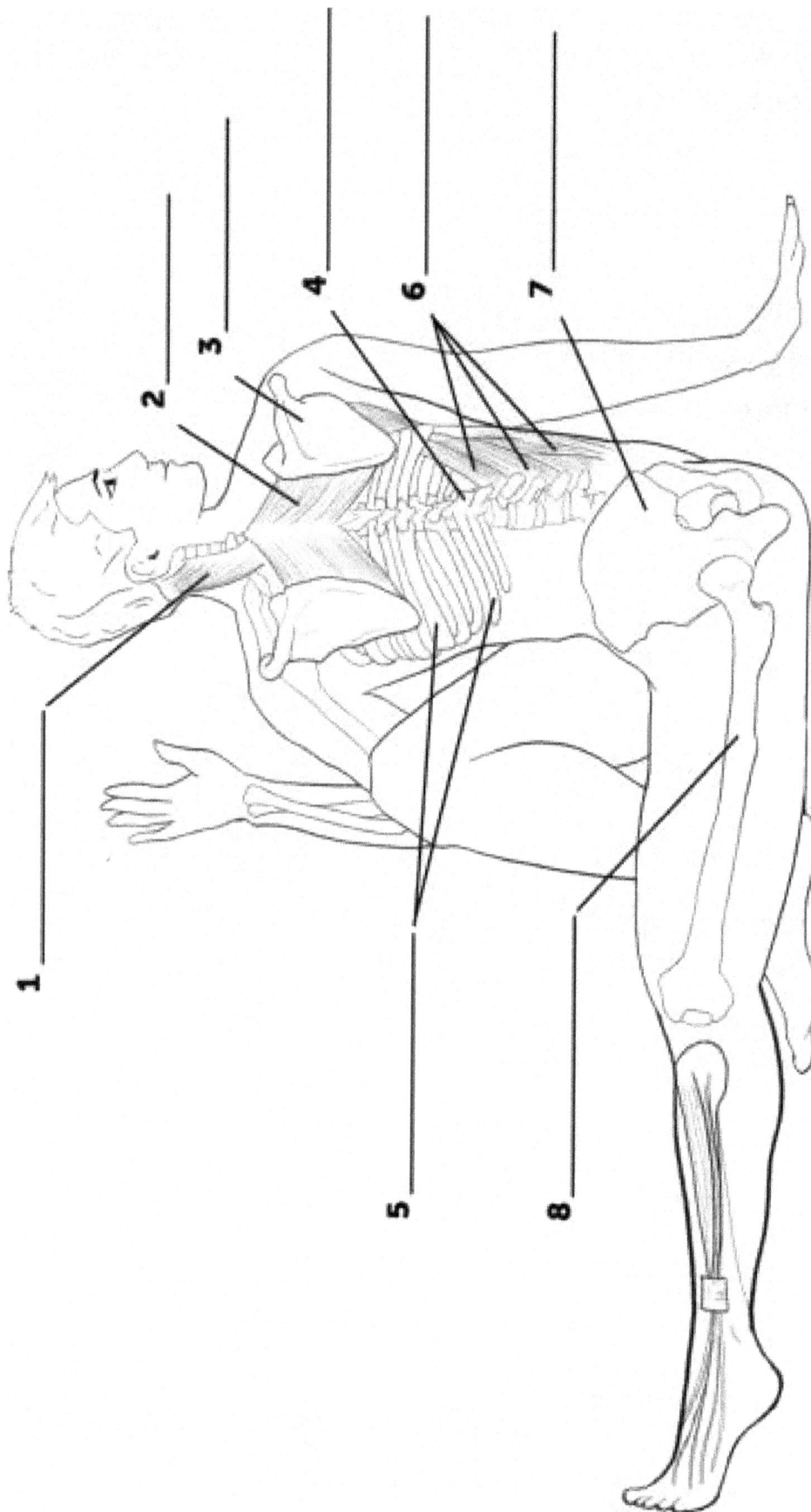

61. MARICHI'S POSE III

1. SPLENIUS CAPITIS
2. RHOMBOIDS
3. SCAPULA
4. SPINE
5. RIBS
6. ERECTOR SPINAE
7. PELVIS
8. FEMUR

62. PYRAMID POSE

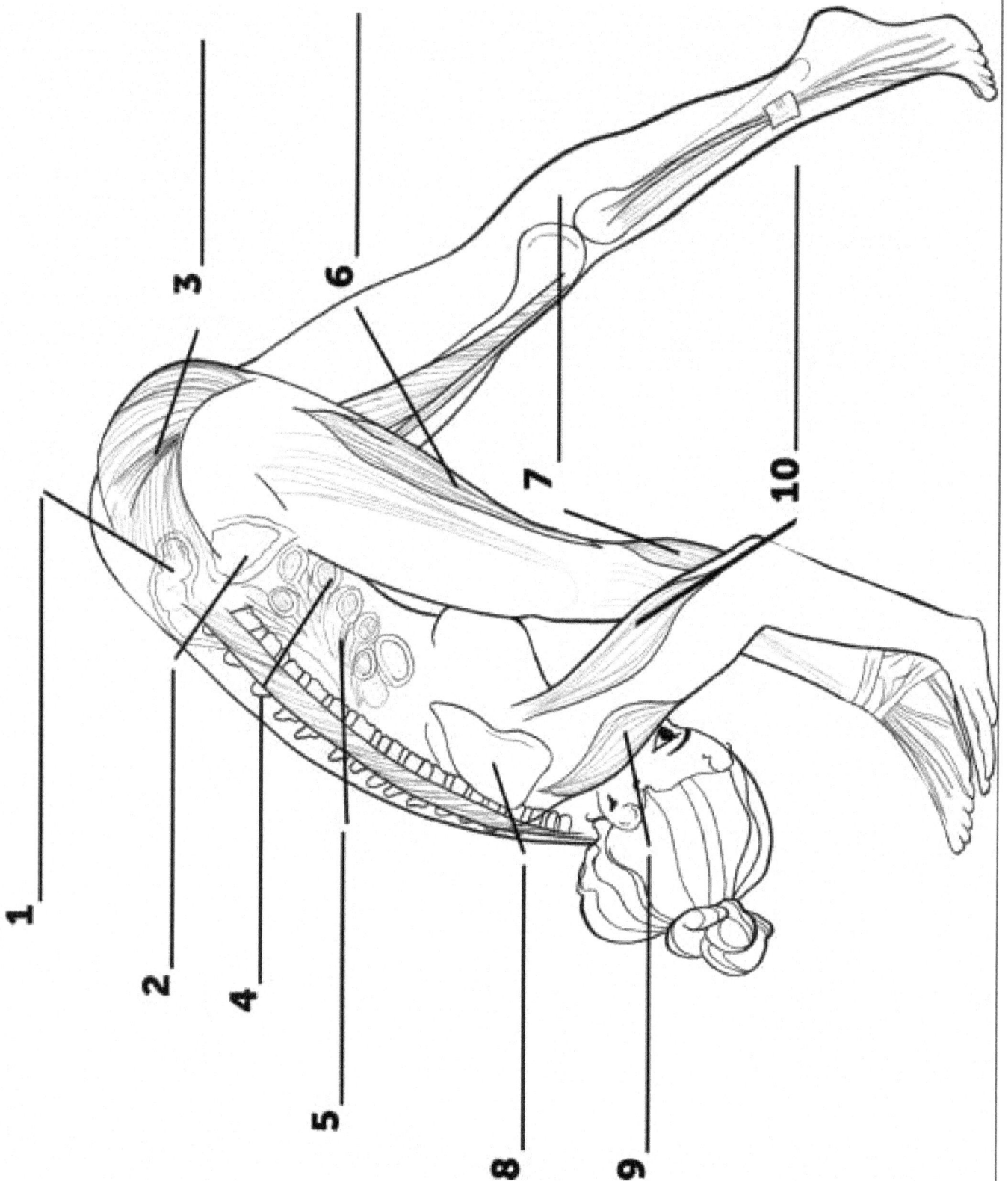

62. PYRAMID POSE

1. RECTUM
2. URINARY BLADDER
3. PIRIFORMIS
4. COILS OF SMALL INTESTINE
5. MESENTERY OF SMALL INTESTINE
6. HAMSTRINGS
7. GASTROCNEMIUS
8. SCAPULA
9. DELTOID
10. TRICEPS BRACHII

63. WARRIOR I POSE

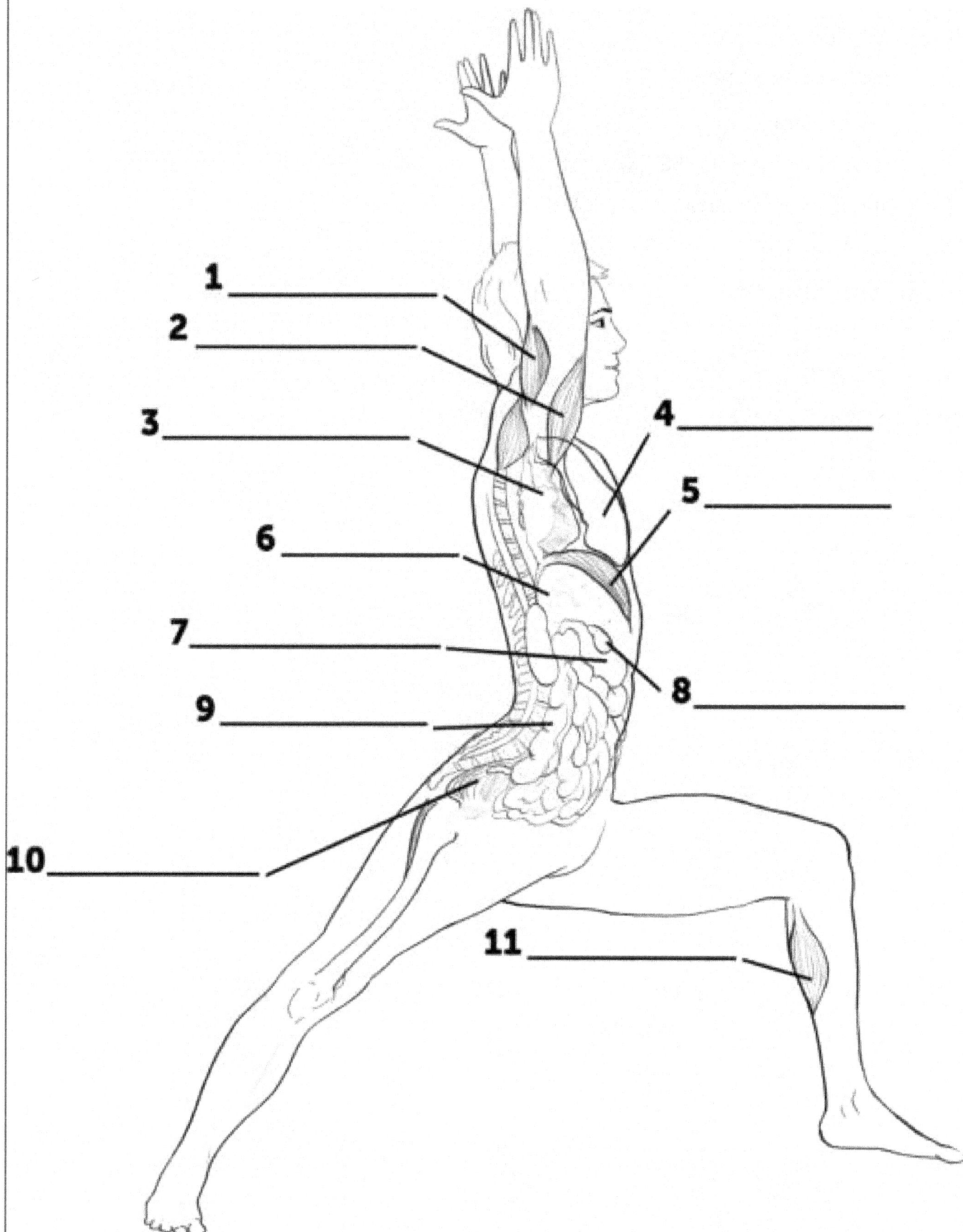

1 _____

2 _____

3 _____

4 _____

5 _____

6 _____

7 _____

8 _____

9 _____

10 _____

11 _____

63. WARRIOR I POSE

1. BICEPS BRACHII
2. TRICEPS BRACHII
3. HEART
4. LUNGS
5. DIAPHRAGM
6. LIVER
7. STOMACH
8. GALLBLADDER
9. ASCENDING COLON
10. RECTUM
11. GASTROCNEMIUS

64. TWISTED WARRIOR POSE

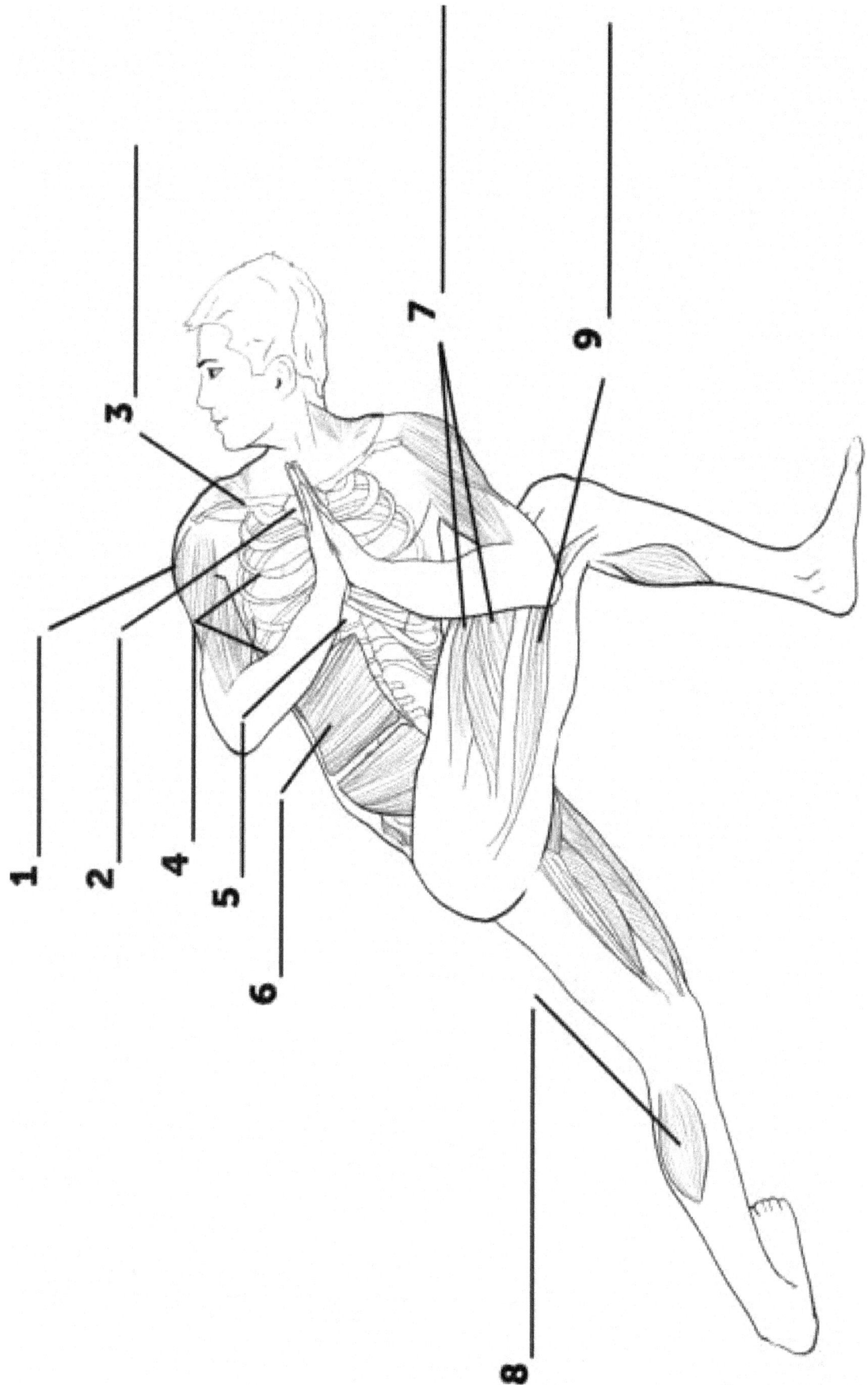

1

2

3

4

5

6

7

8

9

64. TWISTED WARRIOR POSE

1. DELTOID
2. STERNUM
3. COLLARBONE
4. RIBS
5. SPINE
6. INTERNAL OBLIQUE
7. QUADRICEPS
8. GASTROCNEMIUS
9. HAMSTRINGS

65. TWISTED TRIANGLE POSE

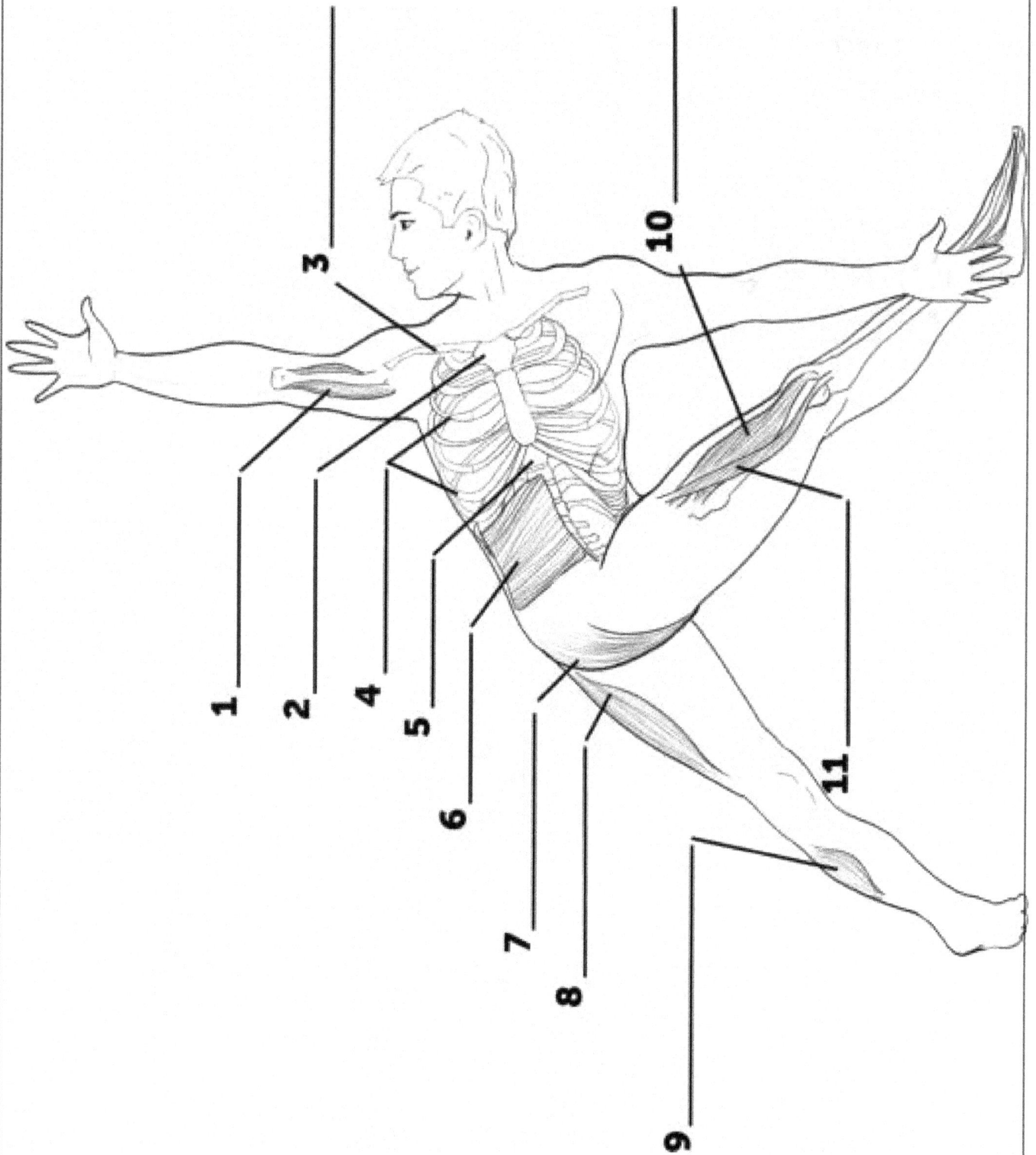

1

2

3

4

5

6

7

8

9

10

11

65. TWISTED TRIANGLE POSE

1.   TRICEPS BRACHII

2.   STERNUM

3.   COLLARBONE

4.   RIBS

5.   SPINE

6.   INTERNAL OBLIQUE

7.   GLUTEUS MAXIMUS

8.   HAMSTRINGS

9.   GASTROCNEMIUS

10. QUADRICEPS

11. SARTORIUS

66. BOUND TWISTED SIDE ANGLE POSE

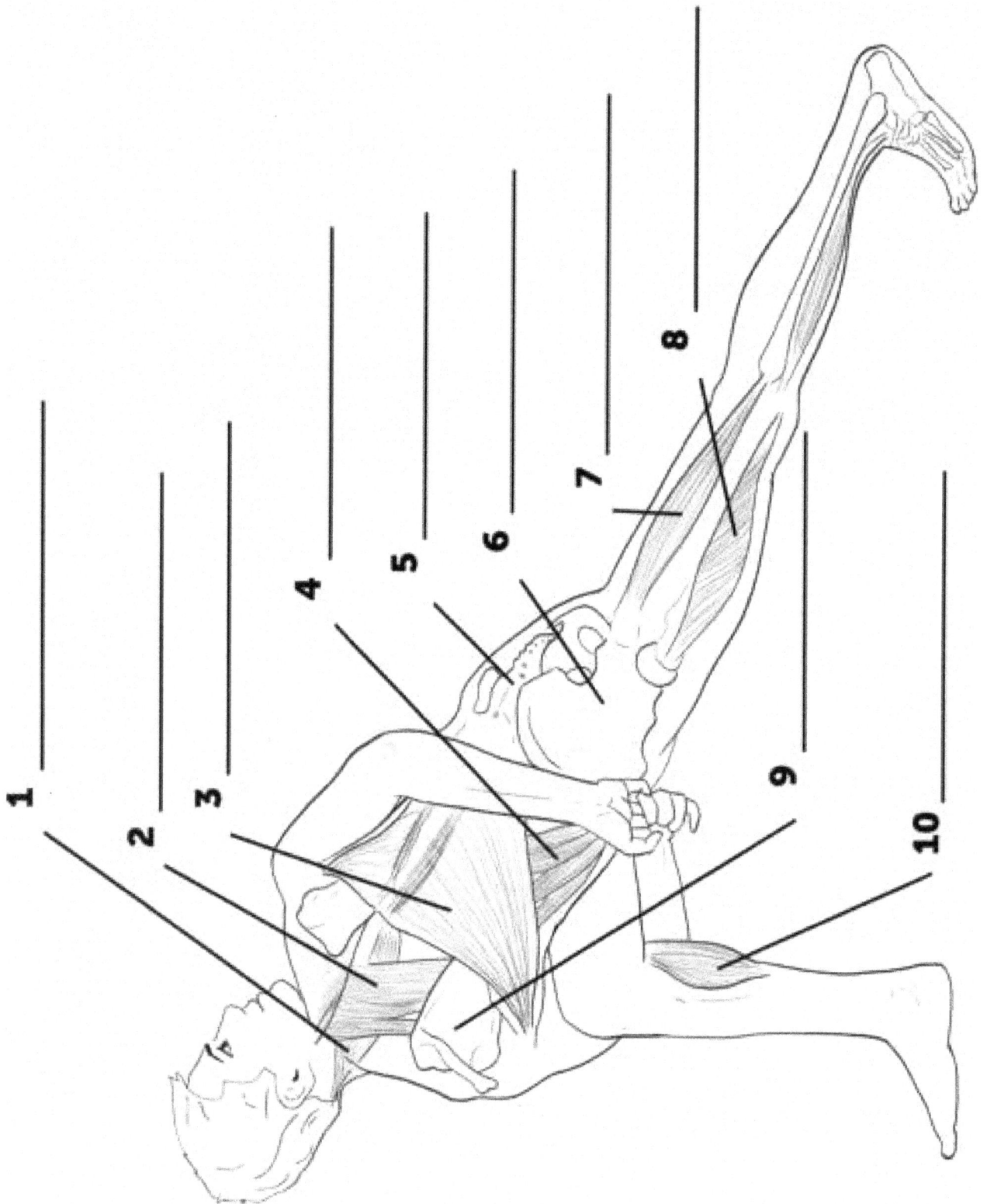

1

2

3

4

5

6

7

8

9

10

66. BOUND TWISTED SIDE ANGLE POSE

1. SPLENIUS CAPITIS

2. RHOMBOIDS

3. LATISSIMUS DORSI

4. ERECTOR SPINAE

5. SACRUM

6. PELVIS

7. HAMSTRINGS

8. QUADRICEPS

9. SCAPULA

10. GASTROCNEMIUS

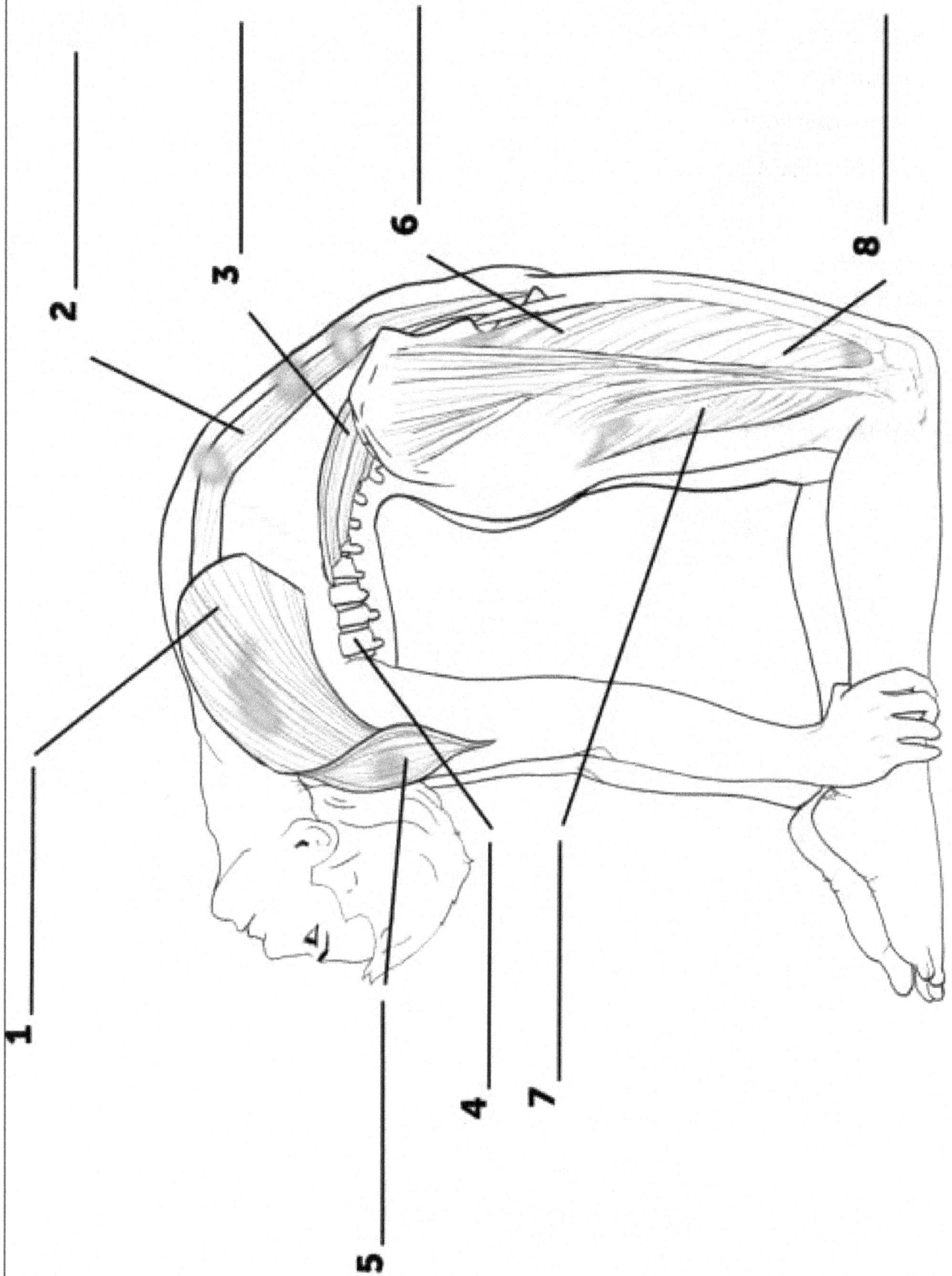

67. CAMEL POSE

1

2

3

4

5

6

7

8

67. CAMEL POSE

1. PECTORALIS MAJOR

2. RECTUS ABDOMINIS

3. PSOAS MAJOR

4. SPINE

5. DELTOID

6. RECTUS FEMORIS

7. HAMSTRINGS

8. VASTUS LATERALIS

68. WARRIOR II POSE

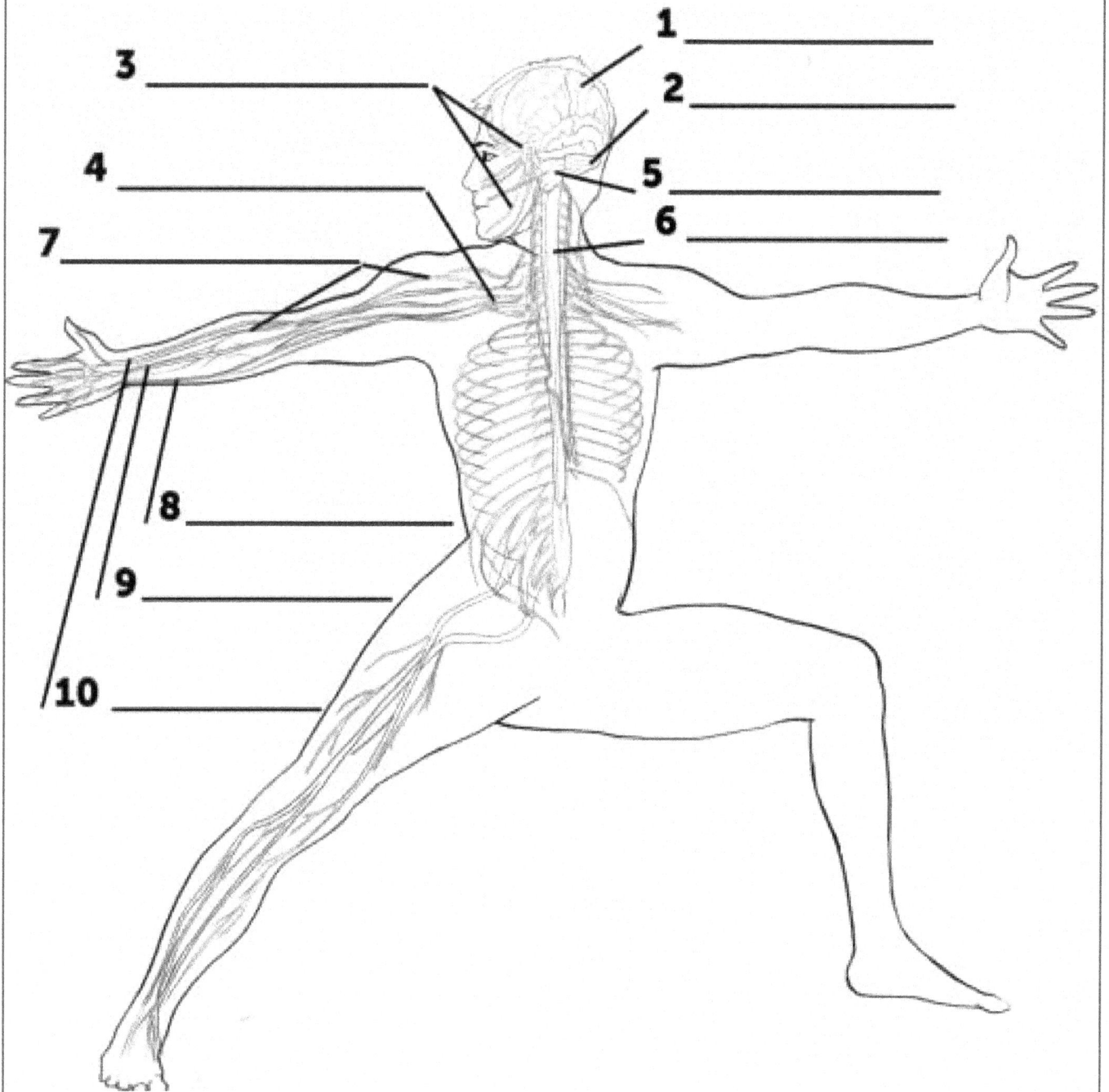

3 _____

4 _____

7 _____

1 _____

2 _____

5 _____

6 _____

8 _____

9 _____

10 _____

68. WARRIOR II POSE

1. CEREBRUM
2. CEREBELLUM
3. CRANIAL NERVES
4. BRACHIAL PLEXUS
5. BRAINSTEM
6. SPINAL CORD
7. MUSCULOCUTANEOUS
8. ULNAR
9. MEDIAN
10. RADIAL

69. WARRIOR III POSE

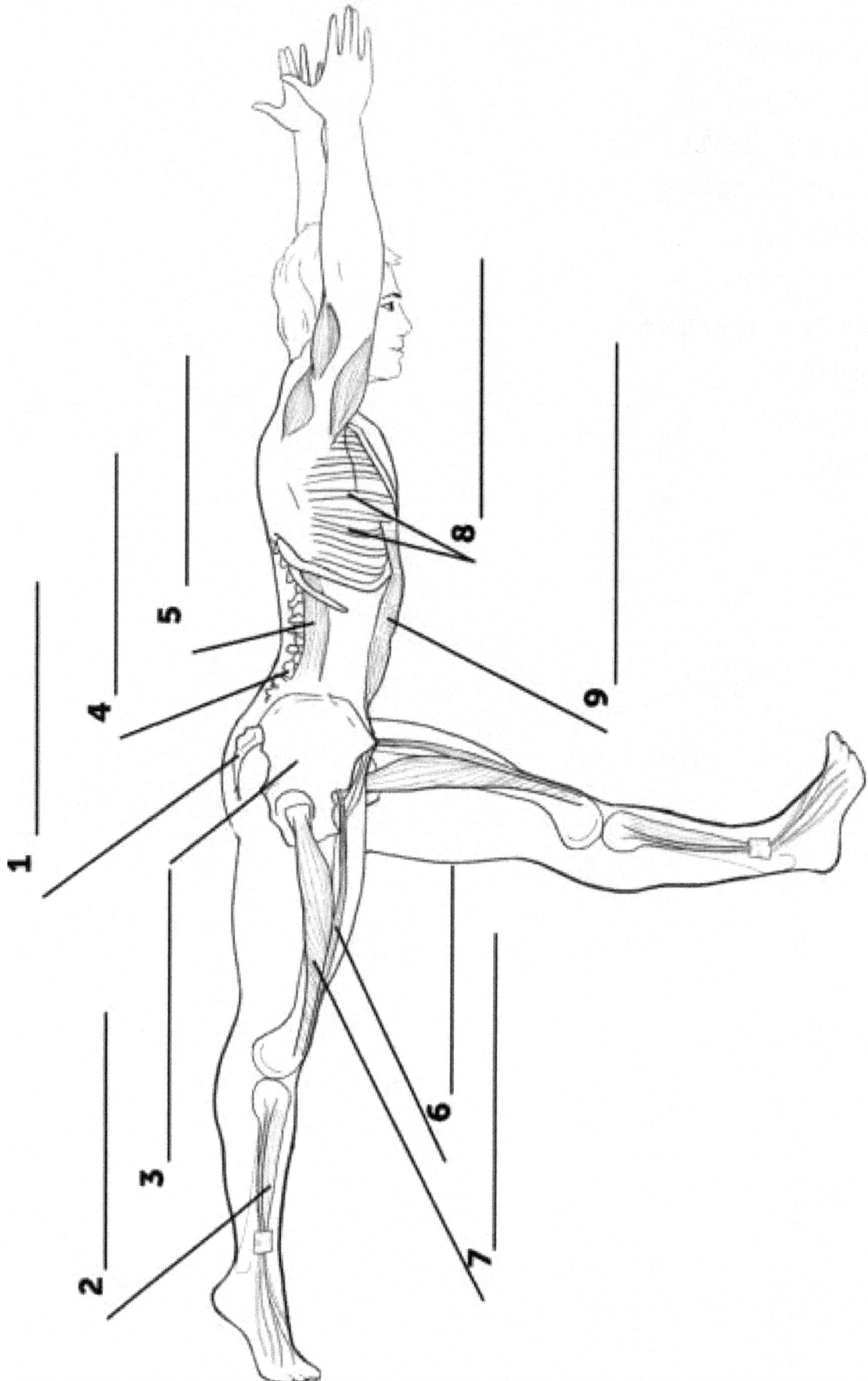

1

2

3

4

5

6

7

8

9

69. WARRIOR III POSE

1.   SACRUM
2.   TIBIALIS ANTERIOR
3.   PELVIS
4.   SPINE
5.   ERECTOR SPINAE
6.   SARTORIUS
7.   RECTUS FEMORIS
8.   RIBS
9.   RECTUS ABDOMINIS

70. REVERSE WARRIOR POSE

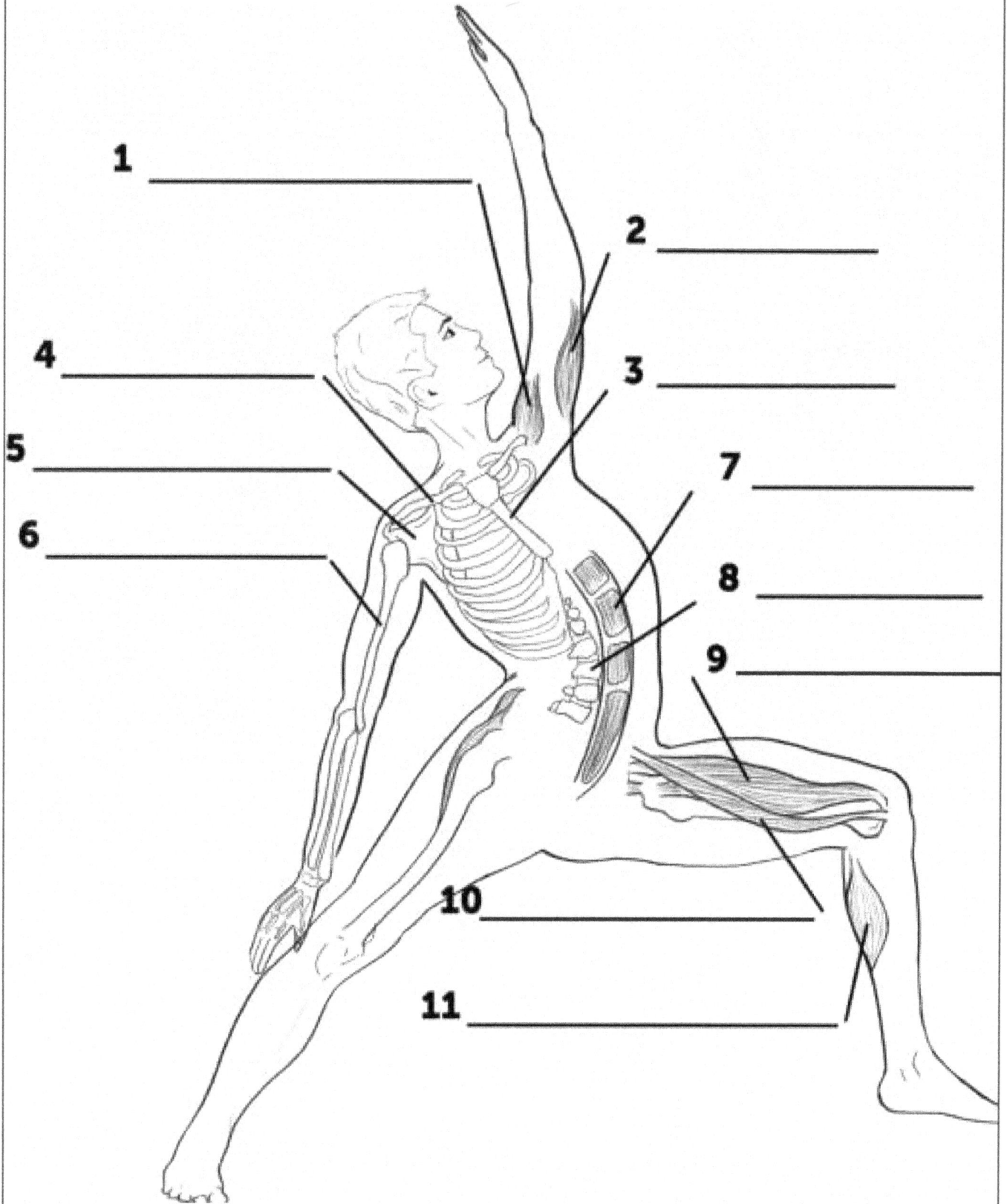

1 _____

2 _____

4 _____

3 _____

5 _____

6 _____

7 _____

8 _____

9 _____

10 _____

11 _____

70. REVERSE WARRIOR POSE

1. DELTOID
2. TRICEPS BRACHII
3. STERNUM
4. COLLARBONE
5. SCAPULA
6. HUMERUS
7. RECTUS ABDOMINIS
8. SPINE
9. RECTUS FEMORIS
10. SARTORIUS
11. GASTROCNEMIUS

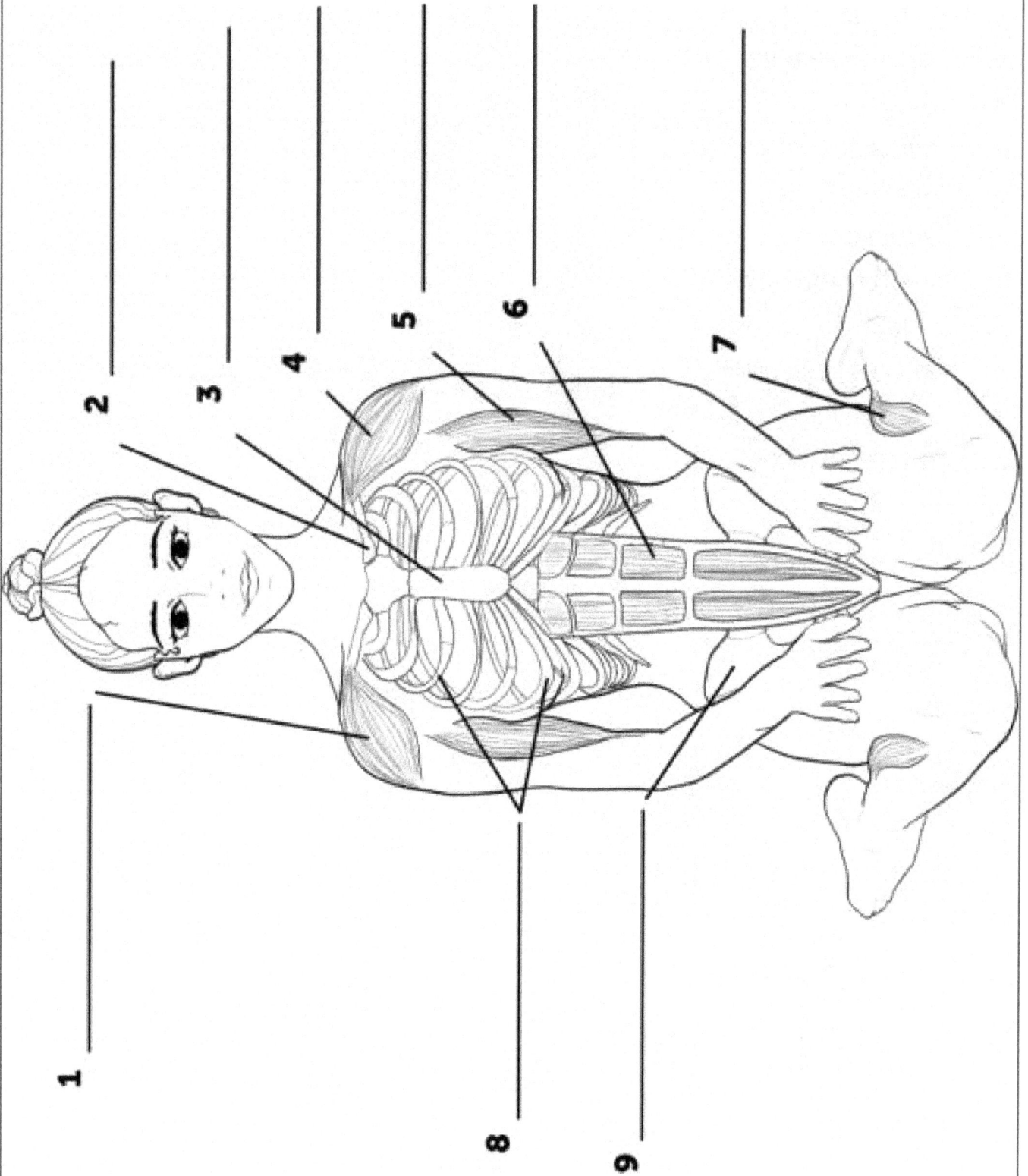

71. HERO POSE

1

2

3

4

5

6

7

8

9

71. HERO POSE

1. DELTOID
2. COLLARBONE
3. STERNUM
4. DELTOID
5. BICEPS BRACHII
6. RECTUS ABDOMINIS
7. GASTROCNEMIUS
8. RIBS
9. PELVIS

72. HALF RECLINED HERO

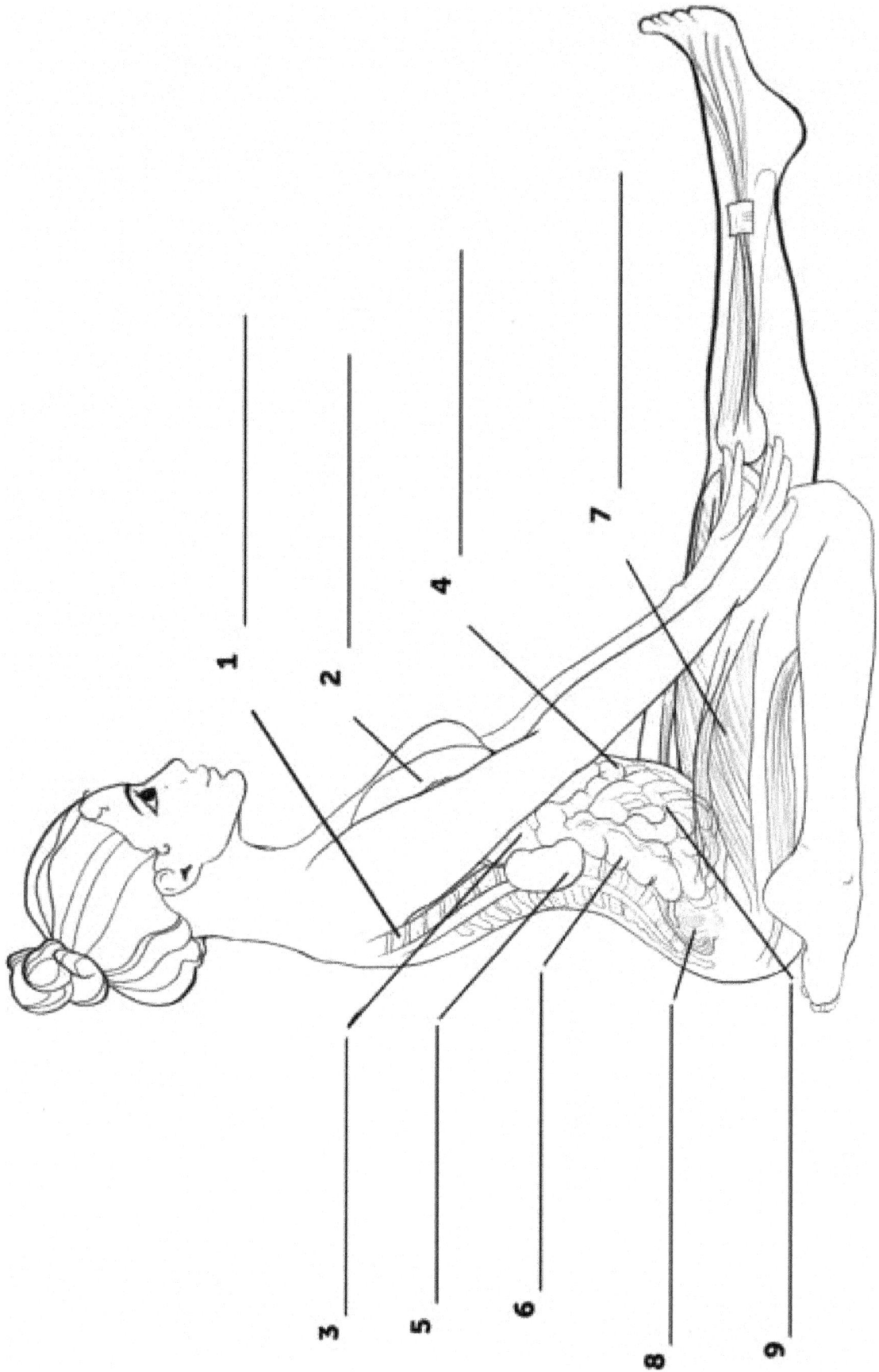

72. HALF RECLINED HERO

1. SPINE
2. LUNGS
3. LIVER
4. TRANSVERSE COLON
5. KIDNEY
6. ASCENDING COLON
7. QUADRICEPS
8. RECTUM
9. COILS OF SMALL INTESTINE

73. RECLINING HERO POSE

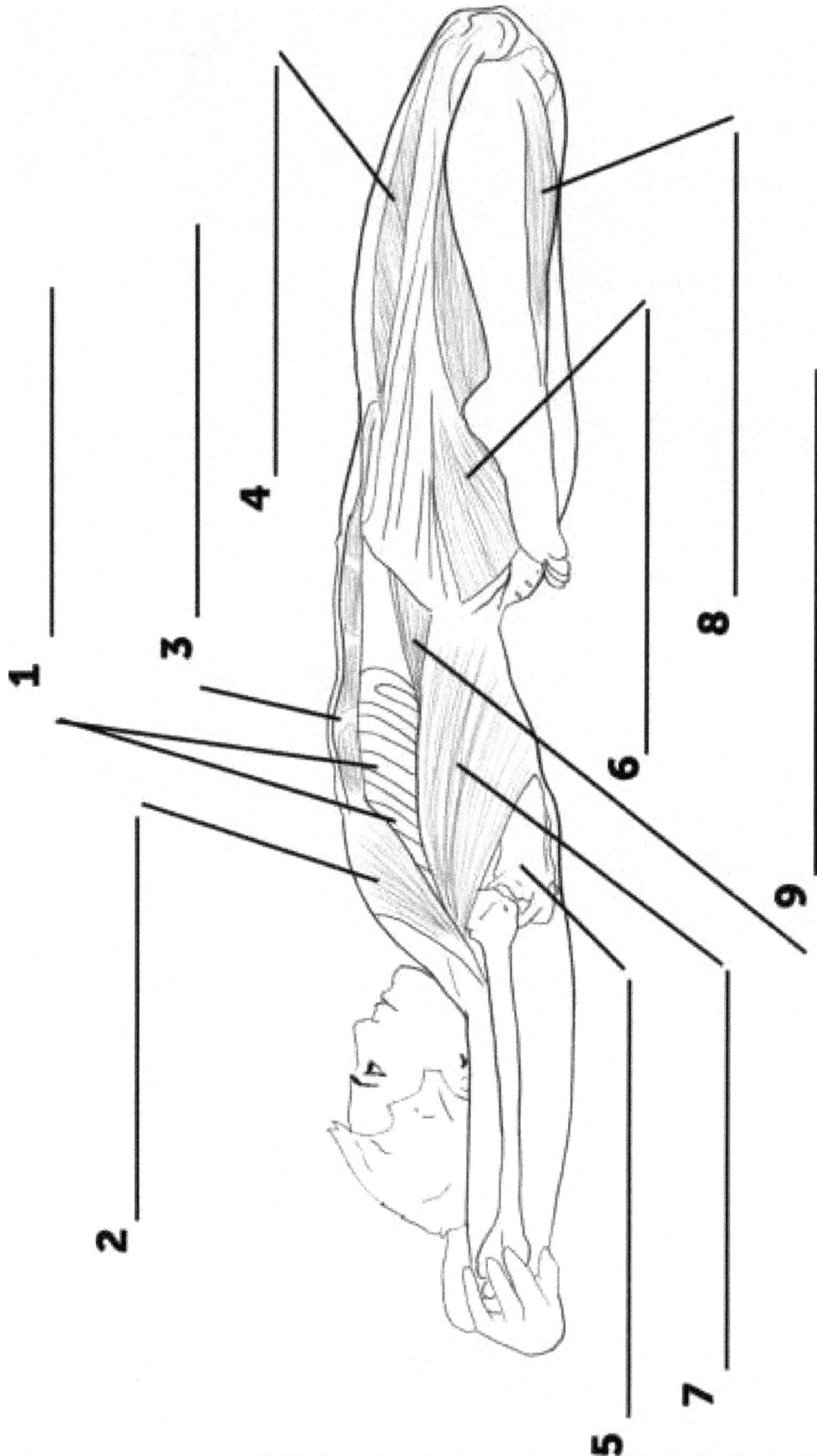

1

2

3

4

5

6

7

8

9

73. RECLINING HERO POSE

1. RIBS
2. PECTORALIS MAJOR
3. RECTUS ABDOMINIS
4. VASTUS LATERALIS
5. SCAPULA
6. GLUTEUS MAXIMUS
7. LATISSIMUS DORSI
8. TIBIALIS ANTERIOR
9. PSOAS MAJOR

74. EXTENDED HAND-TO-BIG-TOE POSE

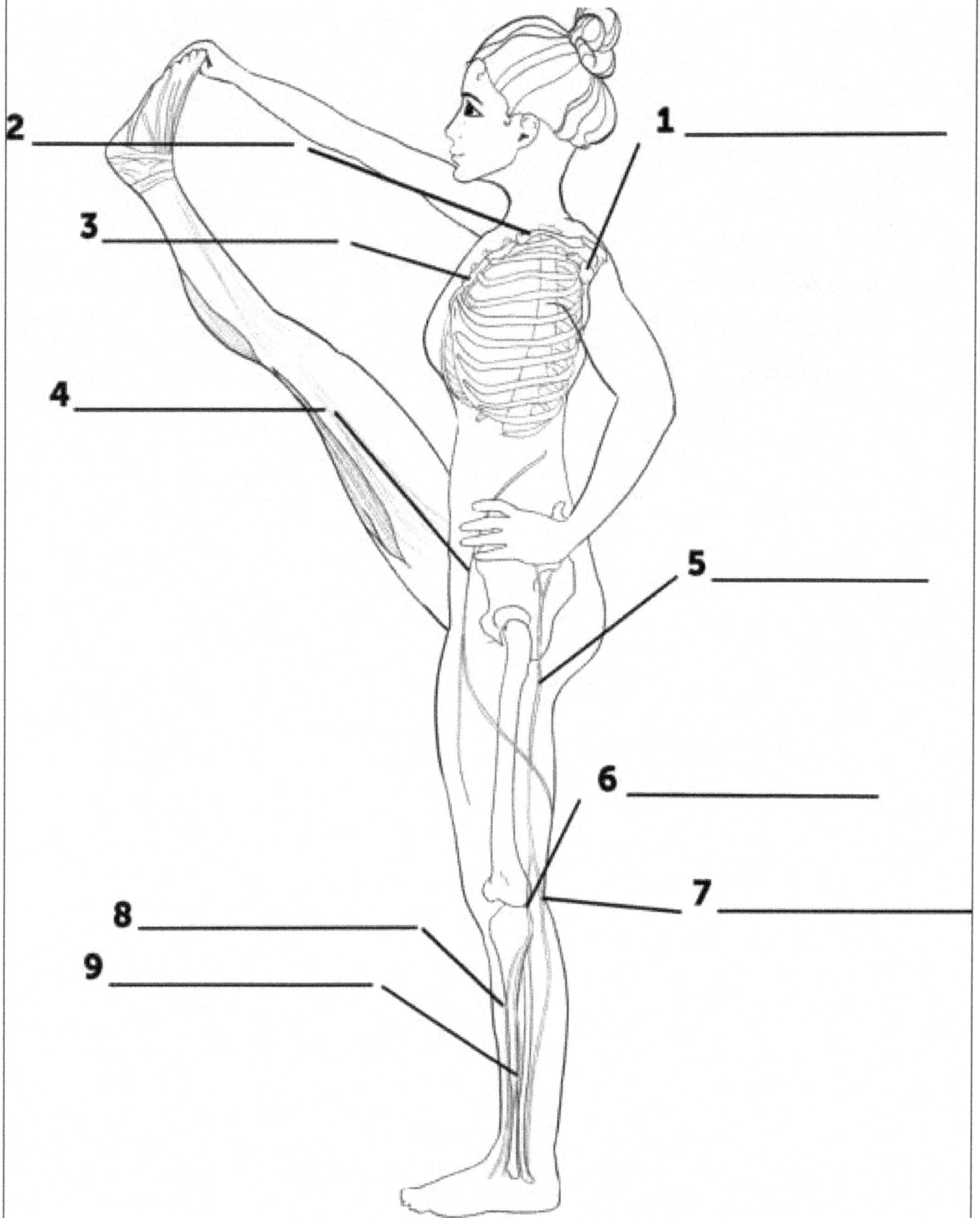

2 _____

3 _____

4 _____

1 _____

5 _____

6 _____

7 _____

8 _____

9 _____

74. EXTENDED HAND-TO-BIG-TOE POSE

1.   SCAPULA

2.   COLLARBONE

3.   STERNUM

4.   LATERAL FEMORAL CUTANEOUS NERVE

5.   SCIATIC NERVE

6.   COMMON PERONEAL NERVE

7.   TIBIAL NERVE

8.   DEEP PERONEAL NERVE

9.   SUPERFICIAL PERONEAL NERVE

75. PIGEON POSE

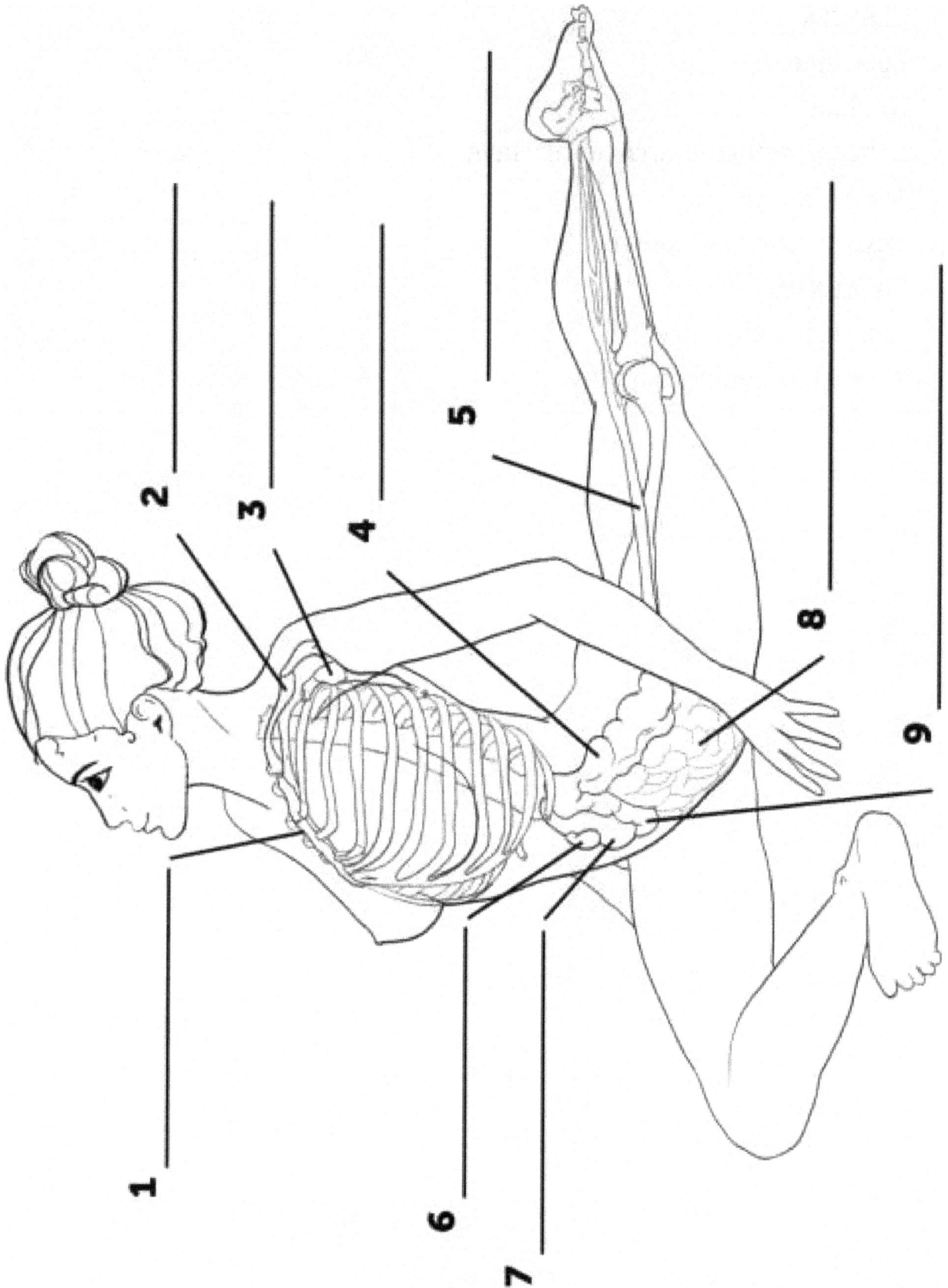

1

2

3

4

5

6

7

8

9

75. PIGEON POSE

1.   STERNUM

2.   COLLARBONE

3.   SCAPULA

4.   ASCENDING COLON

5.   SCIATIC NERVE

6.   GALLBLADDER

7.   STOMACH

8.   COILS OF SMALL INTESTINE

9.   TRANSVERSE COLON

76. THREAD THE NEEDLE

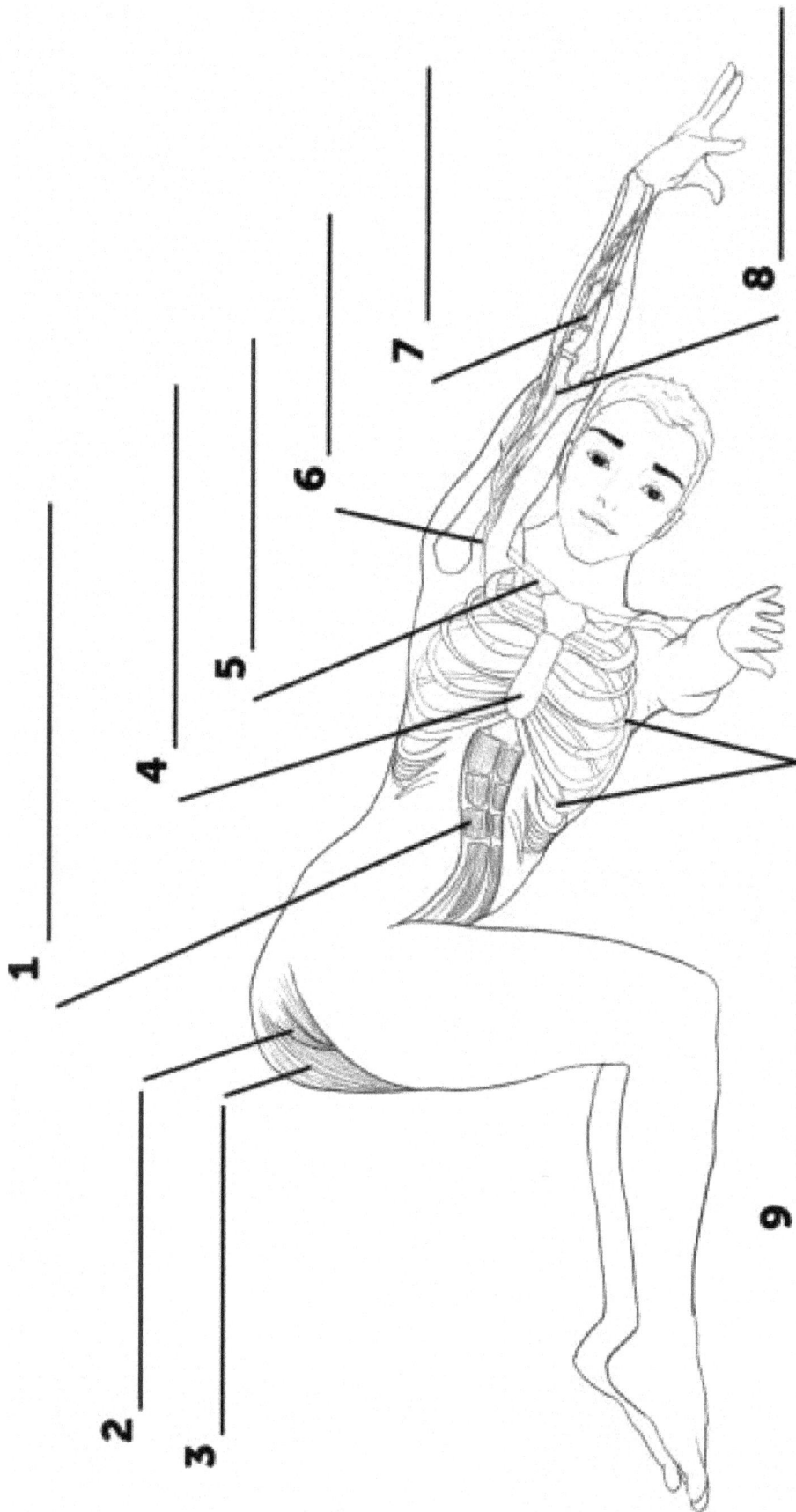

76. THREAD THE NEEDLE

1. RECTUS ABDOMINIS

2. PIRIFORMIS

3. GLUTEUS MAXIMUS

4. STERNUM

5. COLLARBONE

6. RADIAL NERVE

7. POSTERIOR INTEROSSEOUS NERVE

8. ANCONEUS

9. RIBS

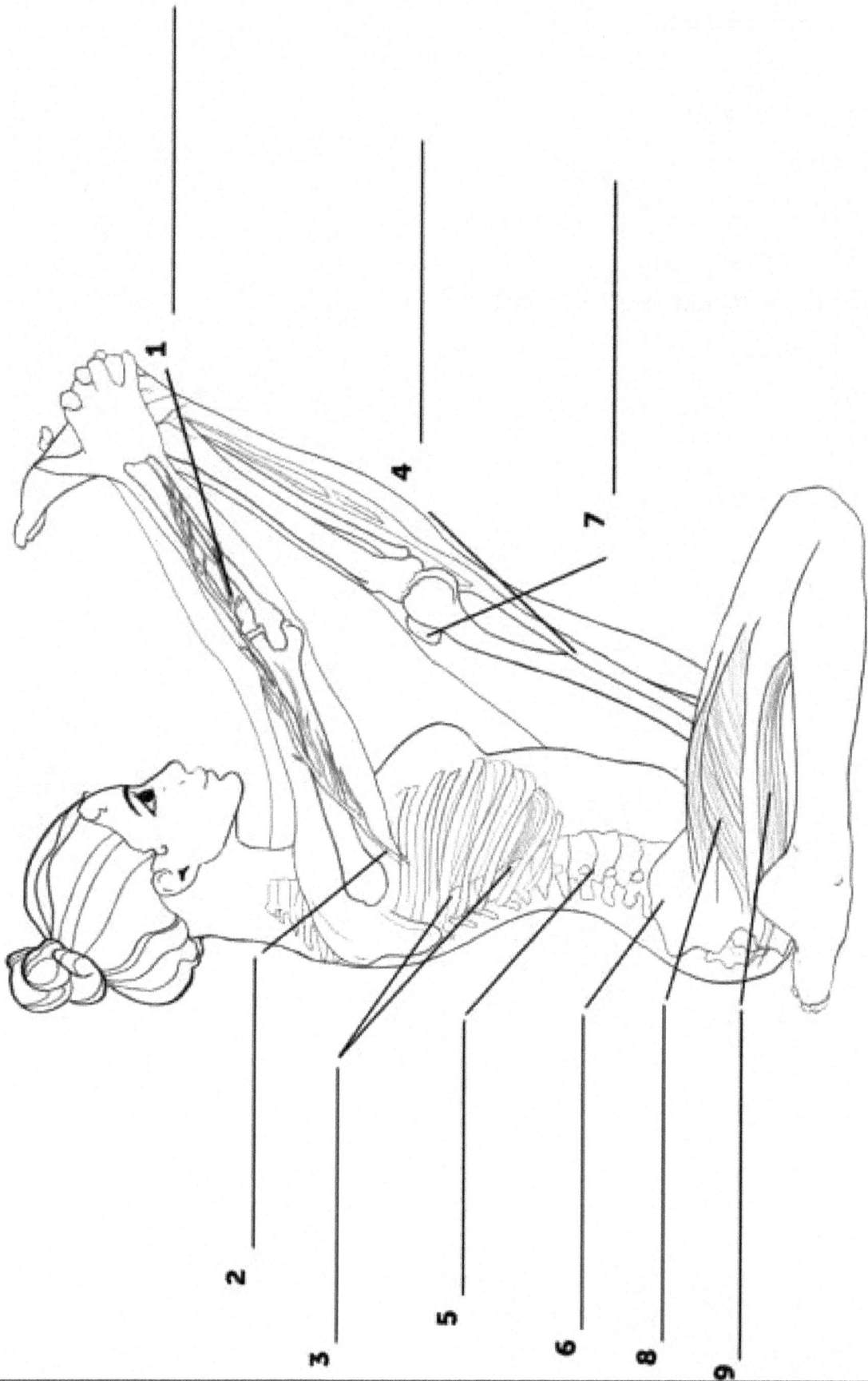

77. HERON POSE

1

2

3

4

5

6

7

8

9

77. HERON POSE

1.   POSTERIOR INTEROSSEOUS NERVE
2.   RADIAL NERVE
3.   RIBS
4.   SCIATIC NERVE
5.   SPINE
6.   PELVIS
7.   PATELLA
8.   QUADRICEPS
9.   HAMSTRINGS

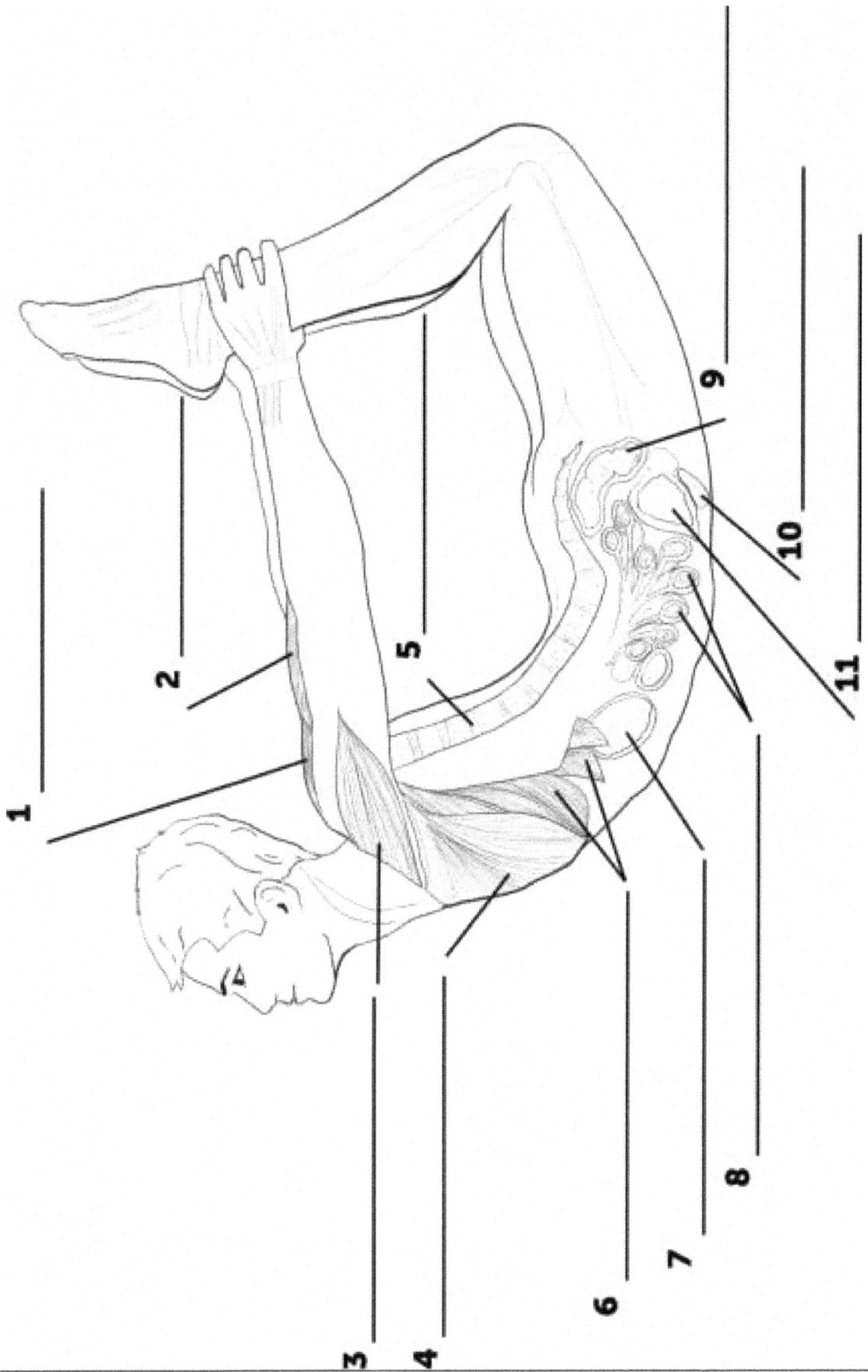

78. BOW POSE

1

2

3

4

5

6

7

8

9

10

11

78. BOW POSE

1. POSTERIOR DELTOID
2. TRICEPS BRACHII
3. ANTERIOR DELTOID
4. PECTORALIS MAJOR
5. SPINE
6. SERRATUS ANTERIOR
7. STOMACH
8. COILS OF SMALL INTESTINE
9. RECTUM
10. PUBIC BONE
11. URINARY BLADDER

79. UPWARD BOW OR WHEEL POSE

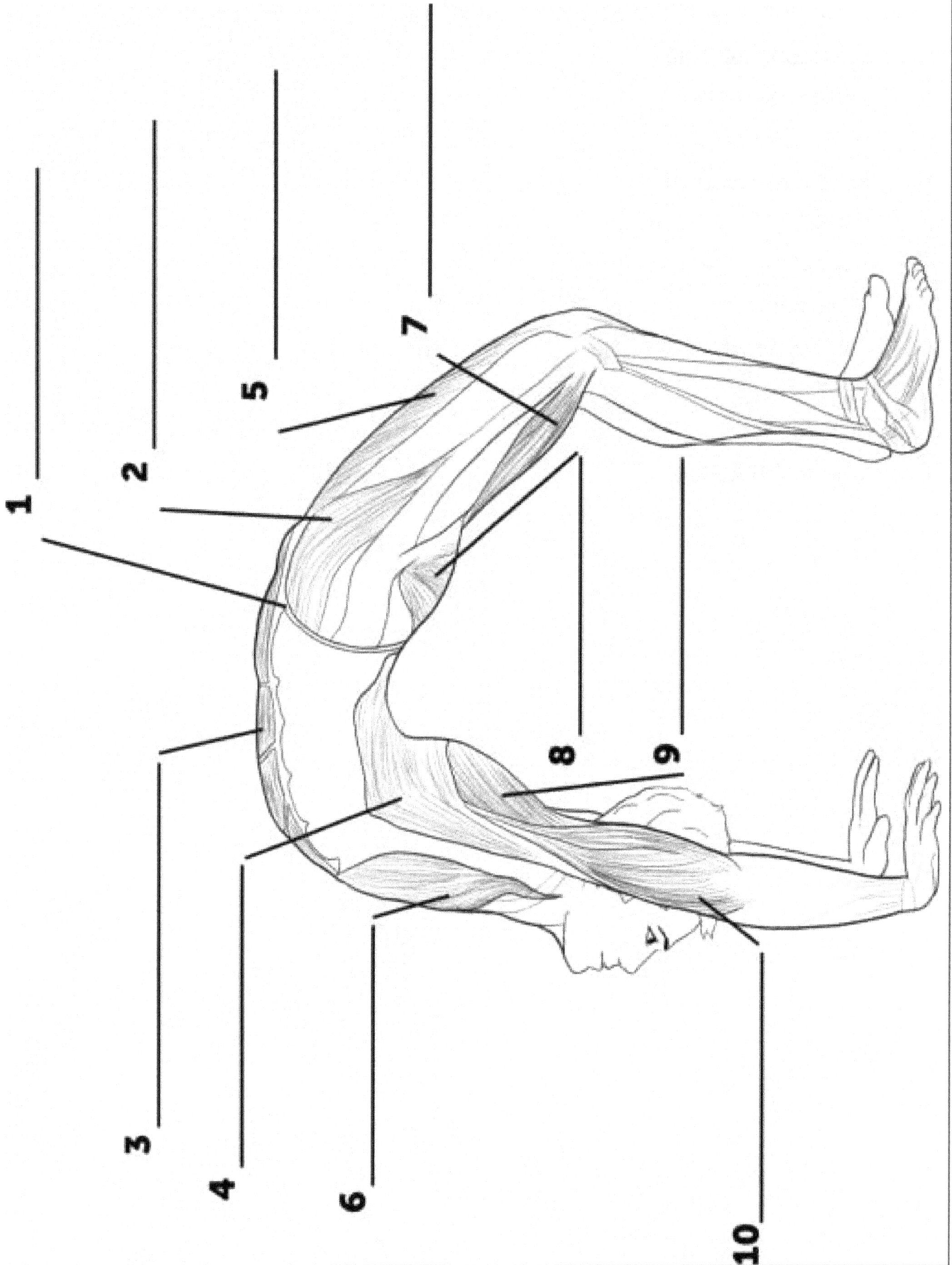

79. UPWARD BOW OR WHEEL POSE

1. ILIOPSOAS
2. TENSOR FASCIA LATA
3. RECTUS ABDOMINIS
4. LATISSIMUS DORSI
5. QUADRICEPS
6. PECTORALIS MAJOR
7. HAMSTRINGS
8. GLUTEUS MAXIMUS
9. ERECTOR SPINAE
10. TRICEPS BRACHII

80. LIZARD POSE

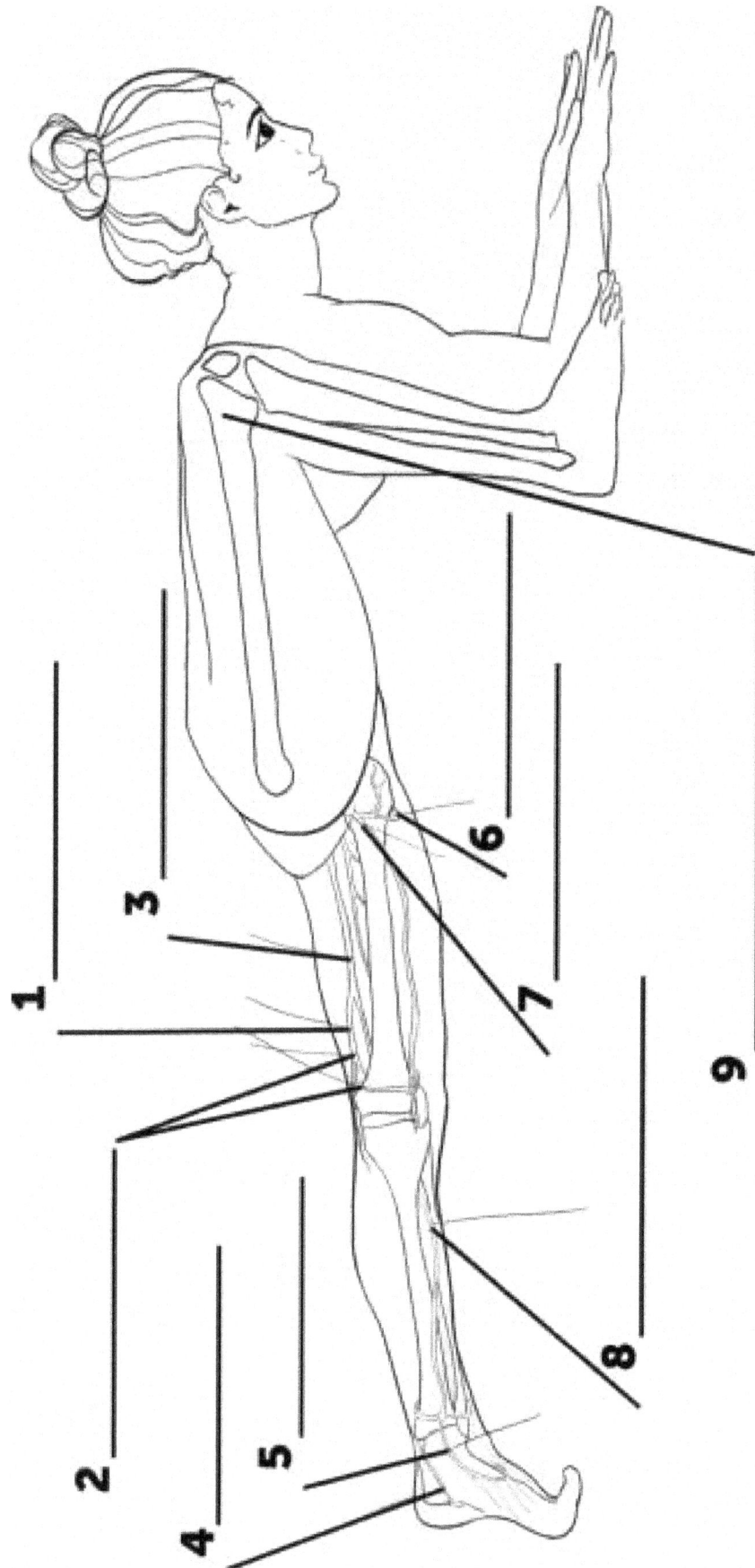

1

2

3

4

5

6

7

8

9

80. LIZARD POSE

1. ADDUCTOR HIATUS
2. GENICULAR ARTERIES
3. FEMORAL ARTERY
4. MEDIAL PLANTAR ARTERY
5. DORSALIS PEDIS ARTERY
6. LATERAL CIRCUMFLEX FEMORAL ARTERY
7. DESCENDING BRANCH
8. ANTERIOR TIBIAL ARTERY
9. FEMUR

81. ONE-LEGGED KING PIGEON POSE

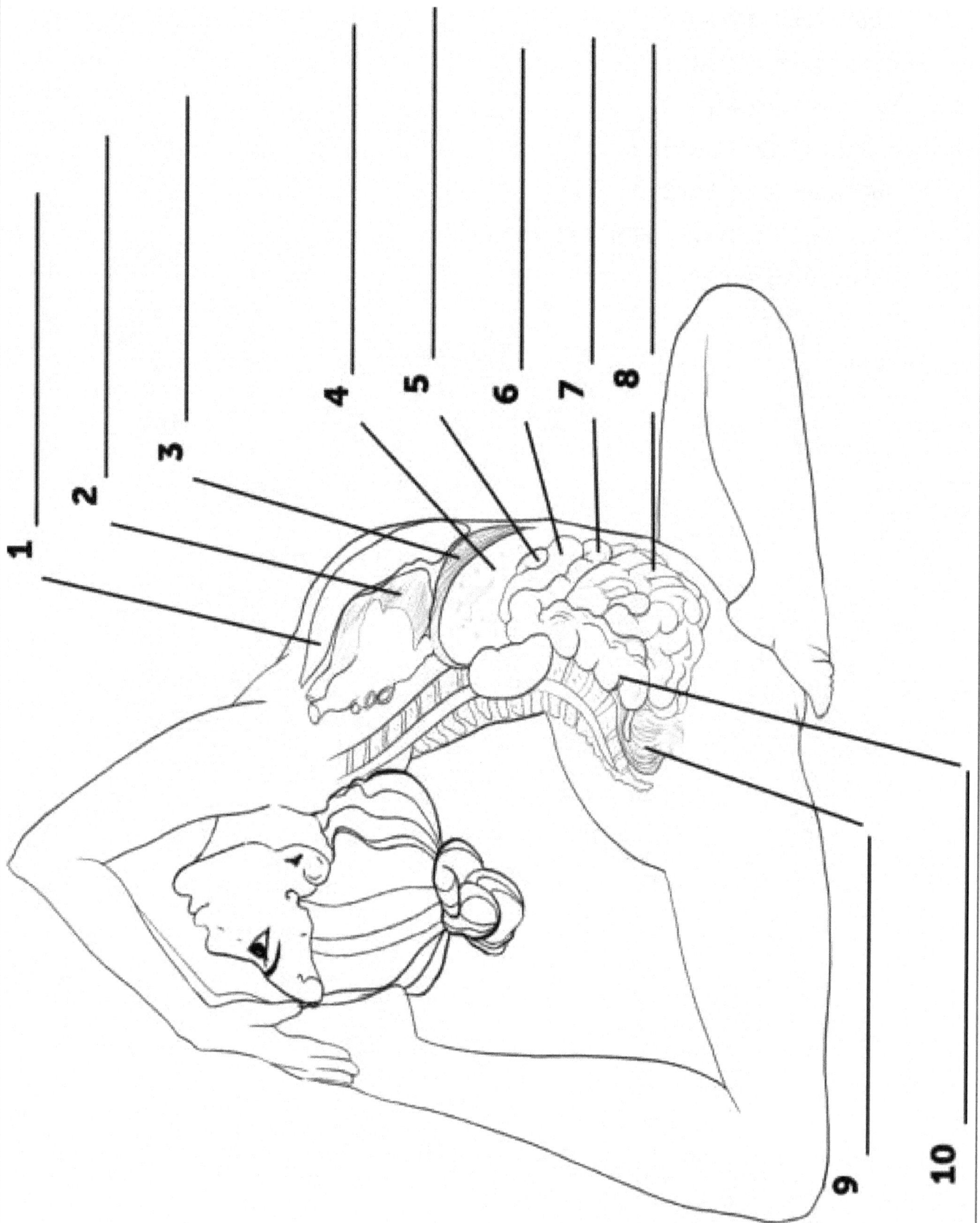

81. ONE-LEGGED KING PIGEON POSE

1. LUNGS
2. HEART
3. DIAPHRAGM
4. LIVER
5. GALLBLADDER
6. STOMACH
7. TRANSVERSE COLON
8. COILS OF SMALL INTESTINE
9. RECTUM
10. ASCENDING COLON

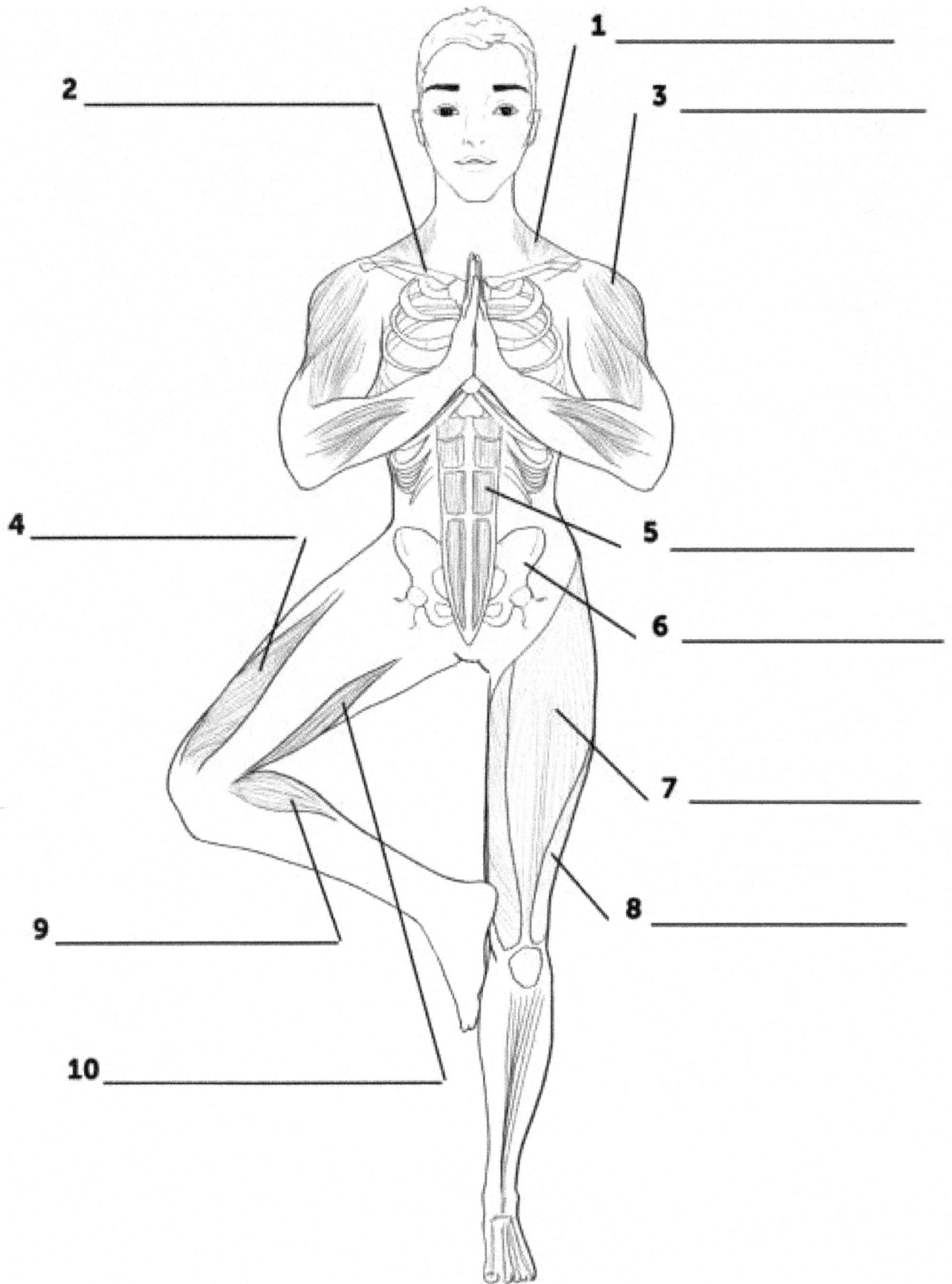

82. TREE POSE

1 _____

2 _____

3 _____

4 _____

5 _____

6 _____

7 _____

8 _____

9 _____

10 _____

82. TREE POSE

1. TRAPEZIUS
2. COLLARBONE
3. DELTOID
4. QUADRICEPS
5. RECTUS ABDOMINIS
6. PELVIS
7. RECTUS FEMORIS
8. VASTUS LATERALIS
9. GASTROCNEMIUS
10. HAMSTRINGS

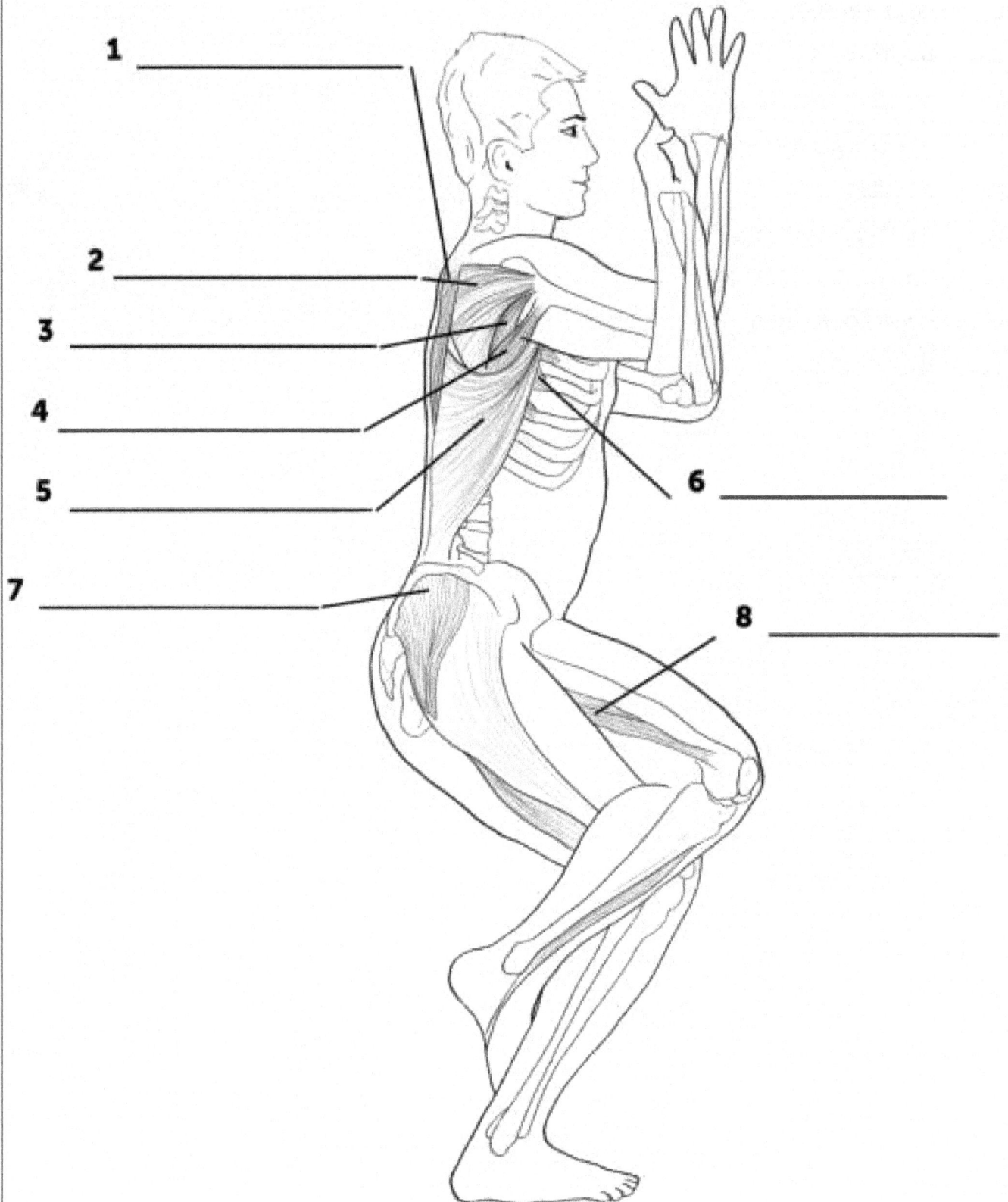

83. EAGLE POSE

1 _____

2 _____

3 _____

4 _____

5 _____

6 _____

7 _____

8 _____

83. EAGLE POSE

1. TRAPEZIUS
2. INFRASPINATUS
3. TERES MINOR
4. TERES MAJOR
5. LATISSIMUS DORSI
6. SERRATUS ANTERIOR
7. GLUTEUS MEDIUS
8. ADDUCTOR MAGNUS

84. HEAD TO KNEE POSE

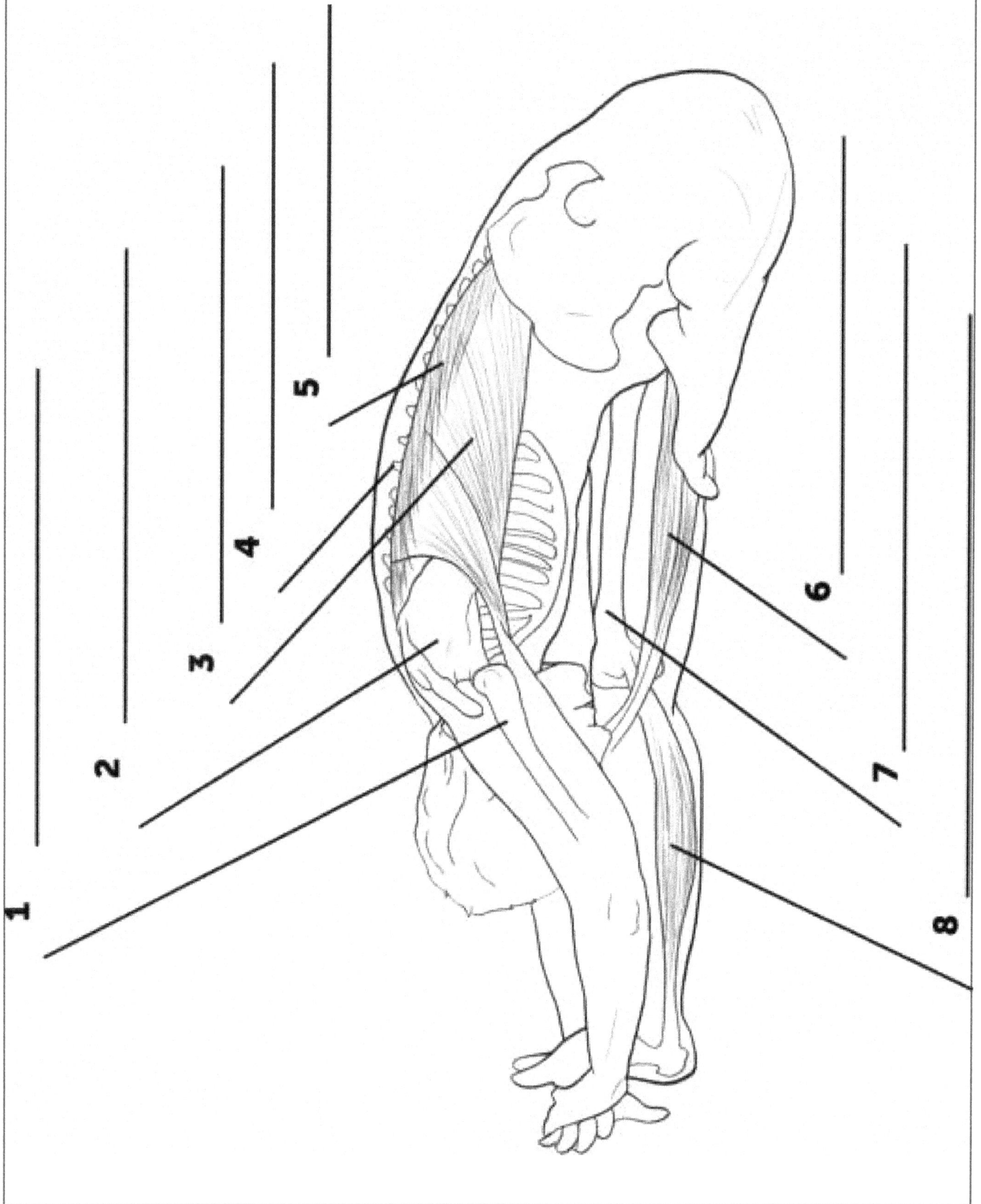

1

2

3

4

5

6

7

8

84. HEAD TO KNEE POSE

1. HUMERUS
2. SCAPULA
3. LATISSIMUS DORSI
4. SPINE
5. ERECTOR SPINAE
6. HAMSTRINGS
7. FEMUR
8. GASTROCNEMIUS

85. LORD OF THE DANCE POSE

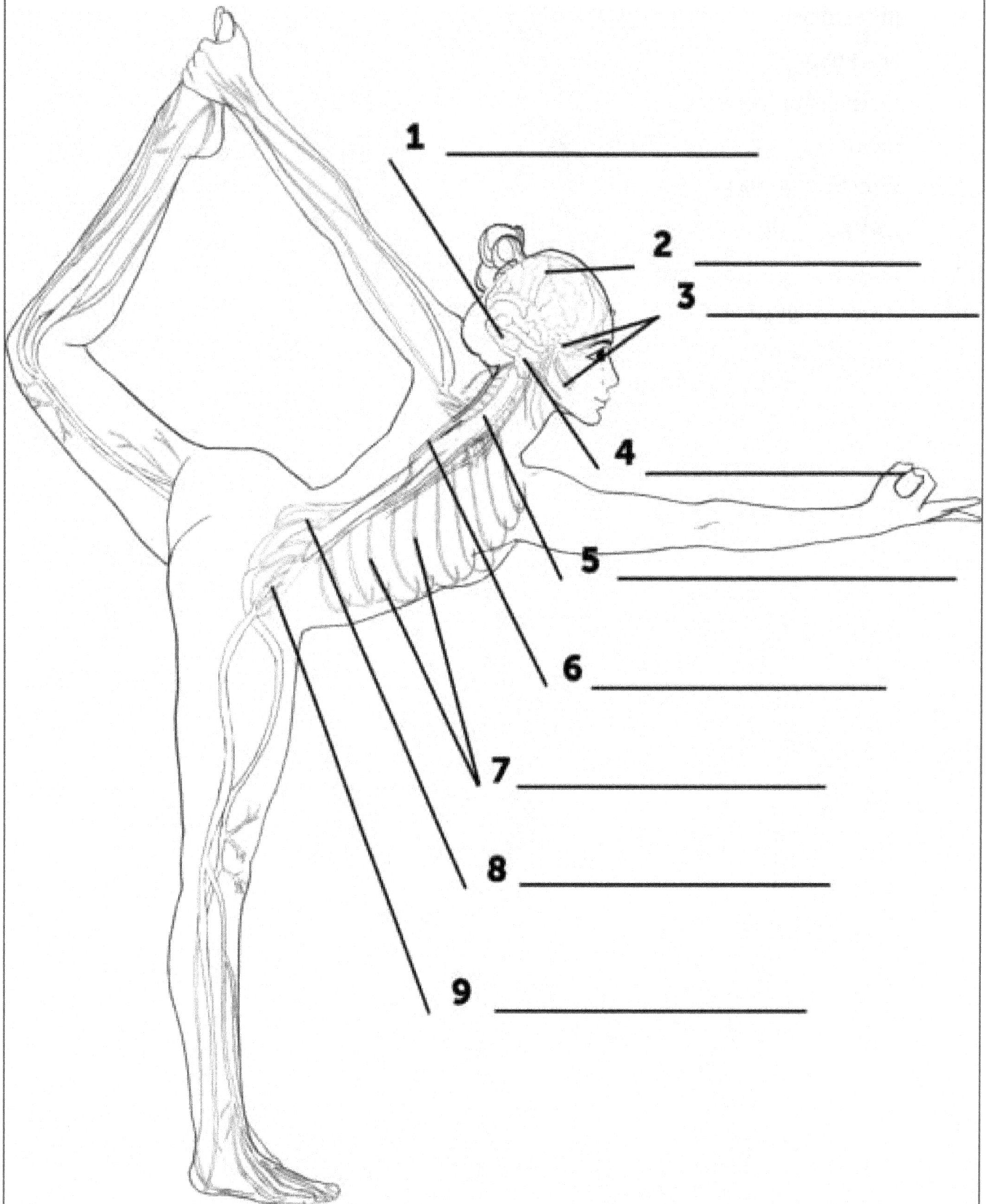

1 _____

2 _____

3 _____

4 _____

5 _____

6 _____

7 _____

8 _____

9 _____

85. LORD OF THE DANCE POSE

1. CEREBELLUM
2. CEREBRUM
3. CRANIAL NERVES
4. BRAINSTEM
5. SPINAL CORD
6. VAGUS
7. INTERCOSTALS
8. LUMBAR PLEXUS
9. SACRAL PLEXUS

86. TWIST CHAIR POSE

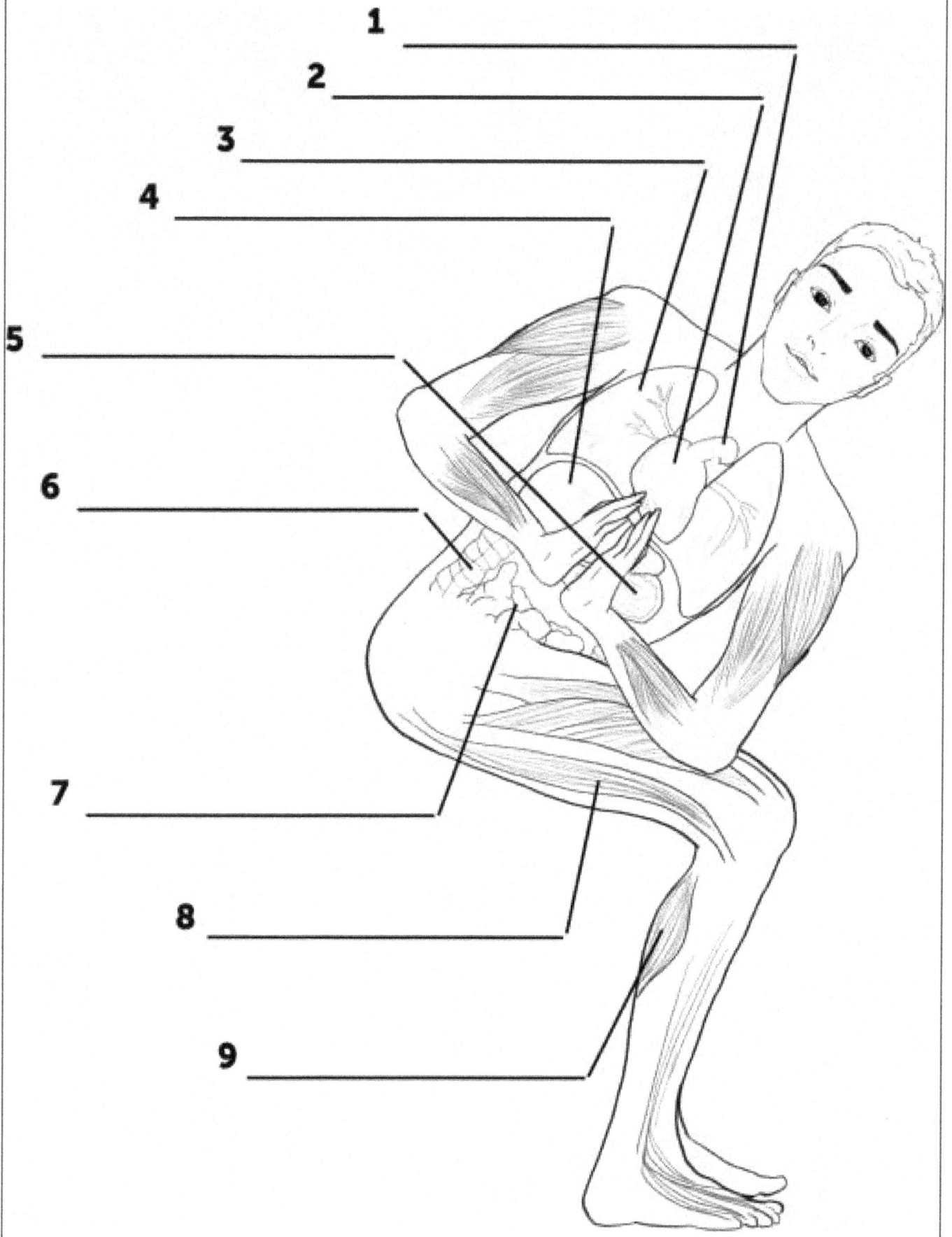

1 _____

2 _____

3 _____

4 _____

5 _____

6 _____

7 _____

8 _____

9 _____

86. TWIST CHAIR POSE

1. AORTA
2. HEART
3. LUNGS
4. LIVER
5. STOMACH
6. ASCENDING COLON
7. COILS OF SMALL INTESTINE
8. HAMSTRINGS
9. GASTROCNEMIUS

87. YOGA RABBIT POSE

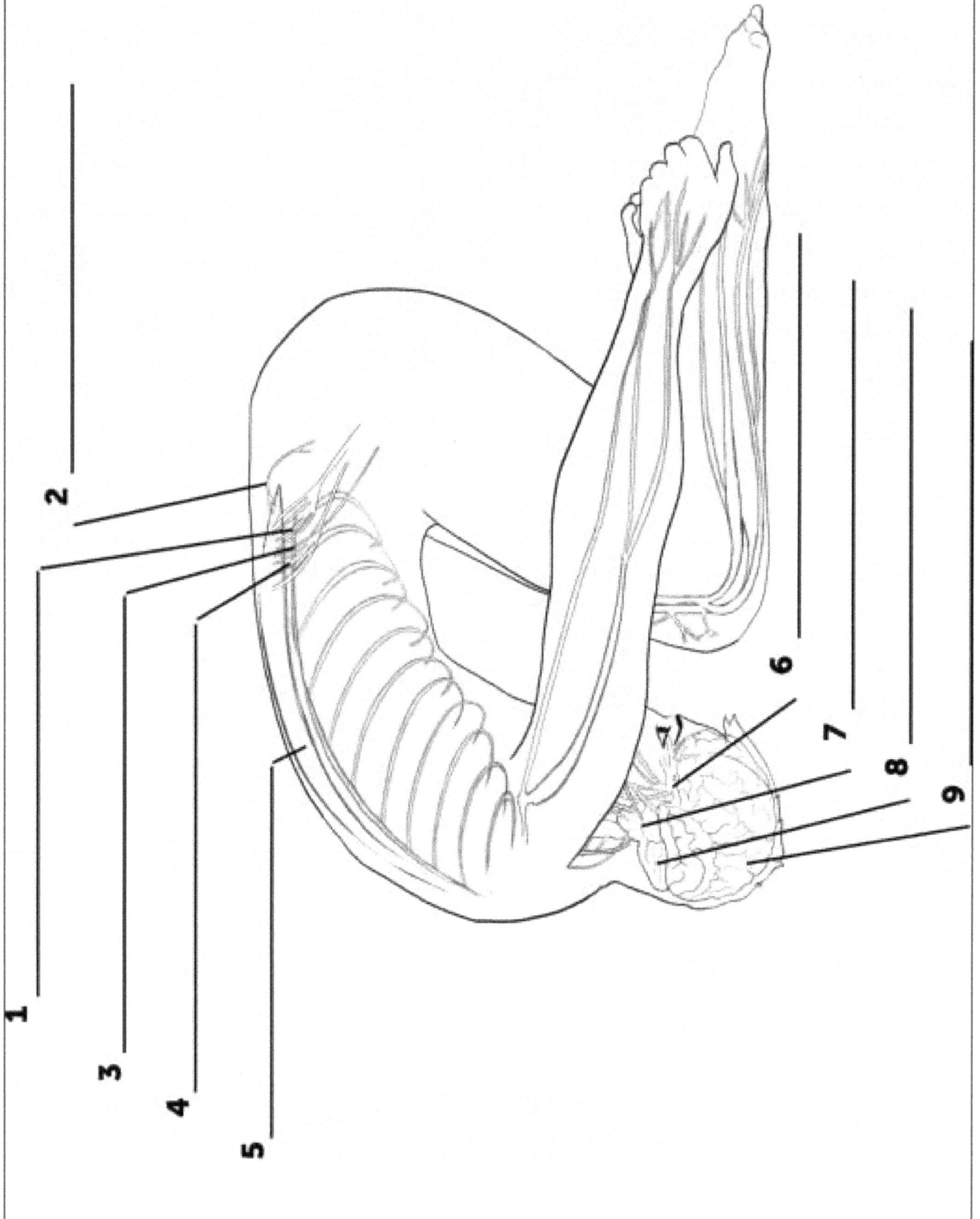

1

2

3

4

5

6

7

8

9

87. YOGA RABBIT POSE

1. SACRAL PLEXUS
2. PUDENDAL NERVE
3. OBTURATOR
4. LUMBAR PLEXUS
5. SPINAL CORD
6. CRANIAL NERVES
7. BRAINSTEM
8. CEREBELLUM
9. CEREBRUM

88. UPWARD PLANK POSE

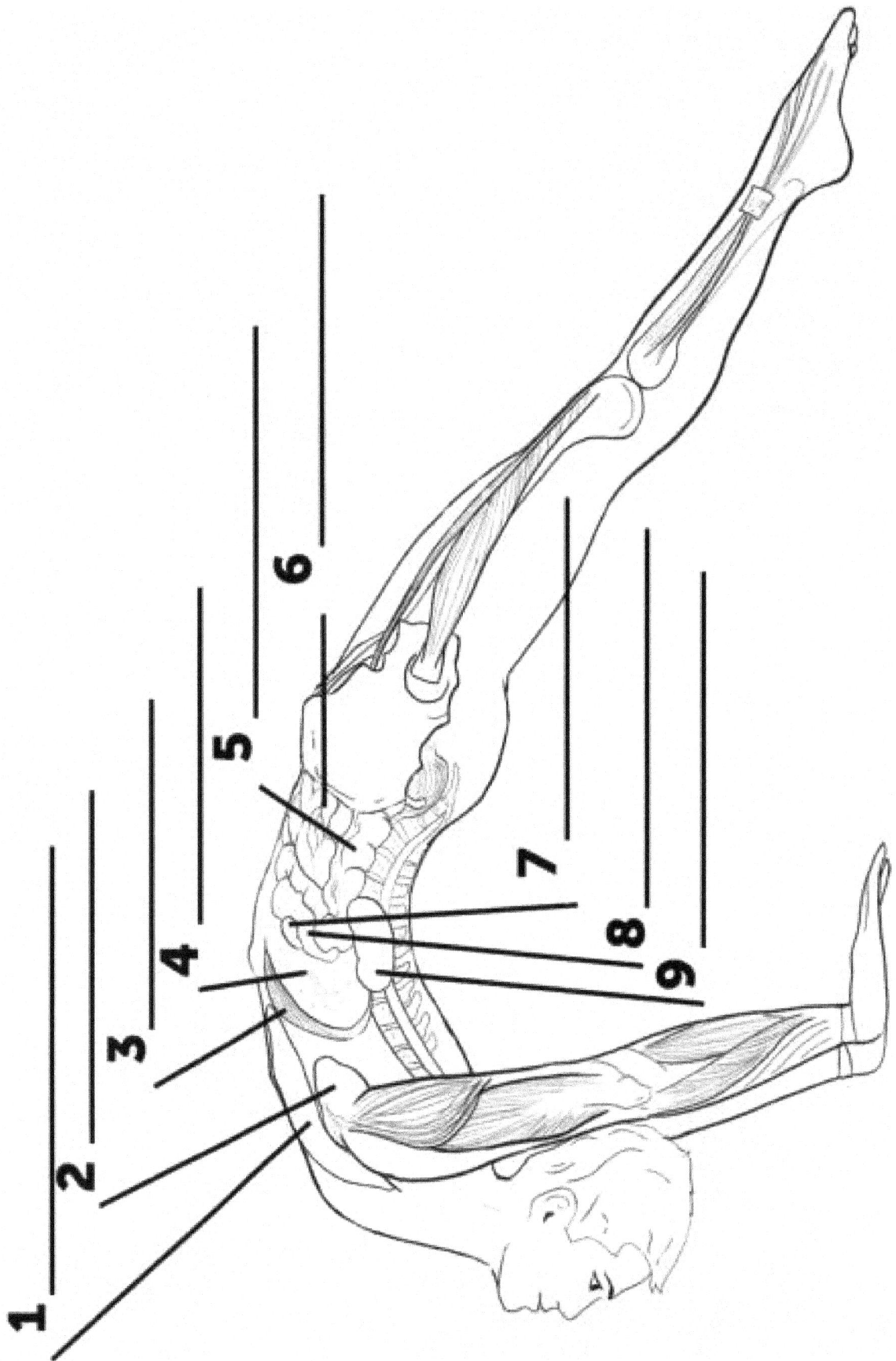

1

2

3

4

5

6

7

8

9

88. UPWARD PLANK POSE

1. LUNGS
2. HEART
3. DIAPHRAGM
4. LIVER
5. ASCENDING COLON
6. COILS OF SMALL INTESTINE
7. GALLBLADDER
8. STOMACH
9. KIDNEY

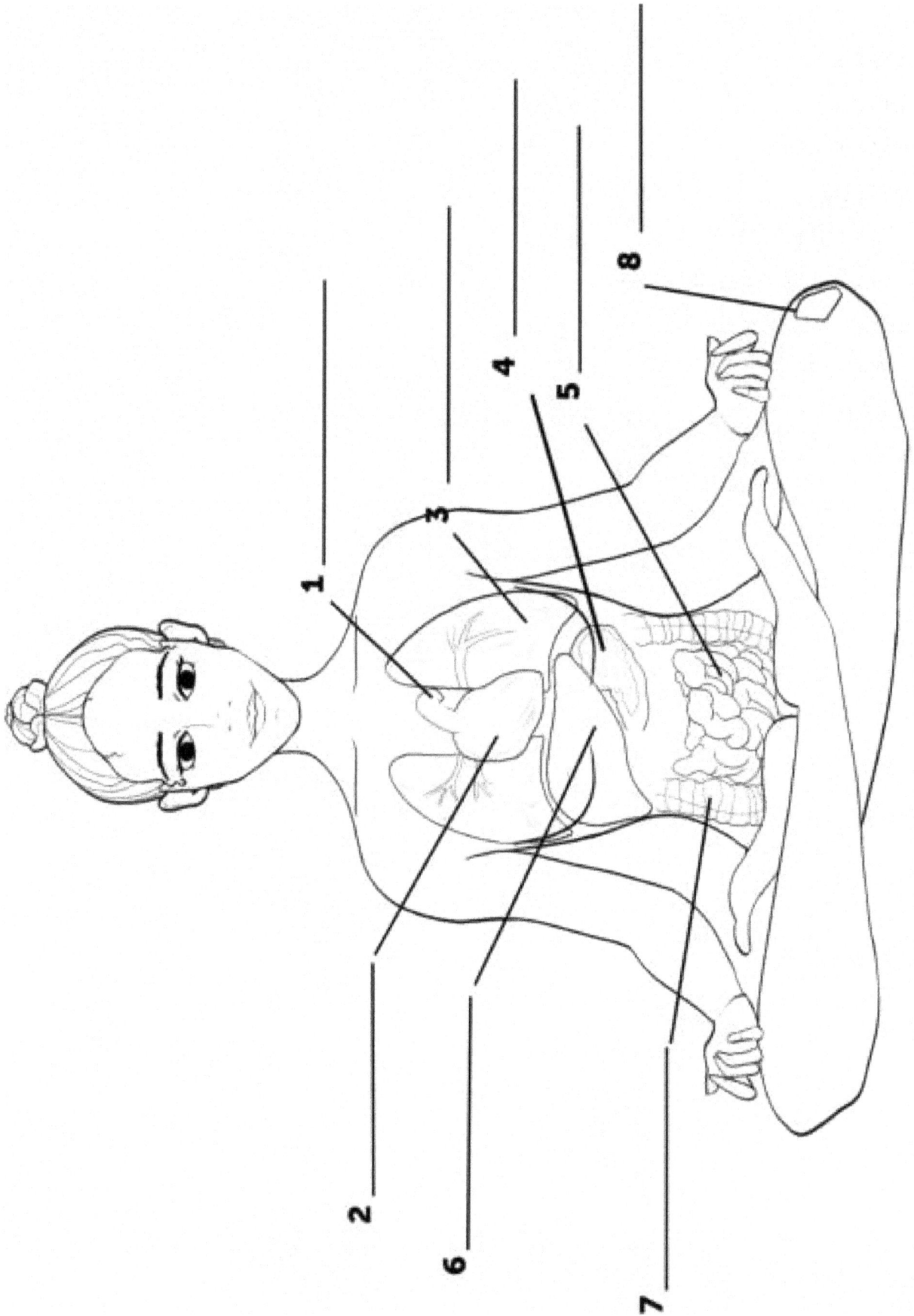

89. LOTUS POSE

1

2

3

4

5

6

7

8

89. LOTUS POSE

1. AORTA
2. HEART
3. LUNGS
4. STOMACH
5. COILS OF SMALL INTESTINE
6. LIVER
7. ASCENDING COLON
8. PATELLA

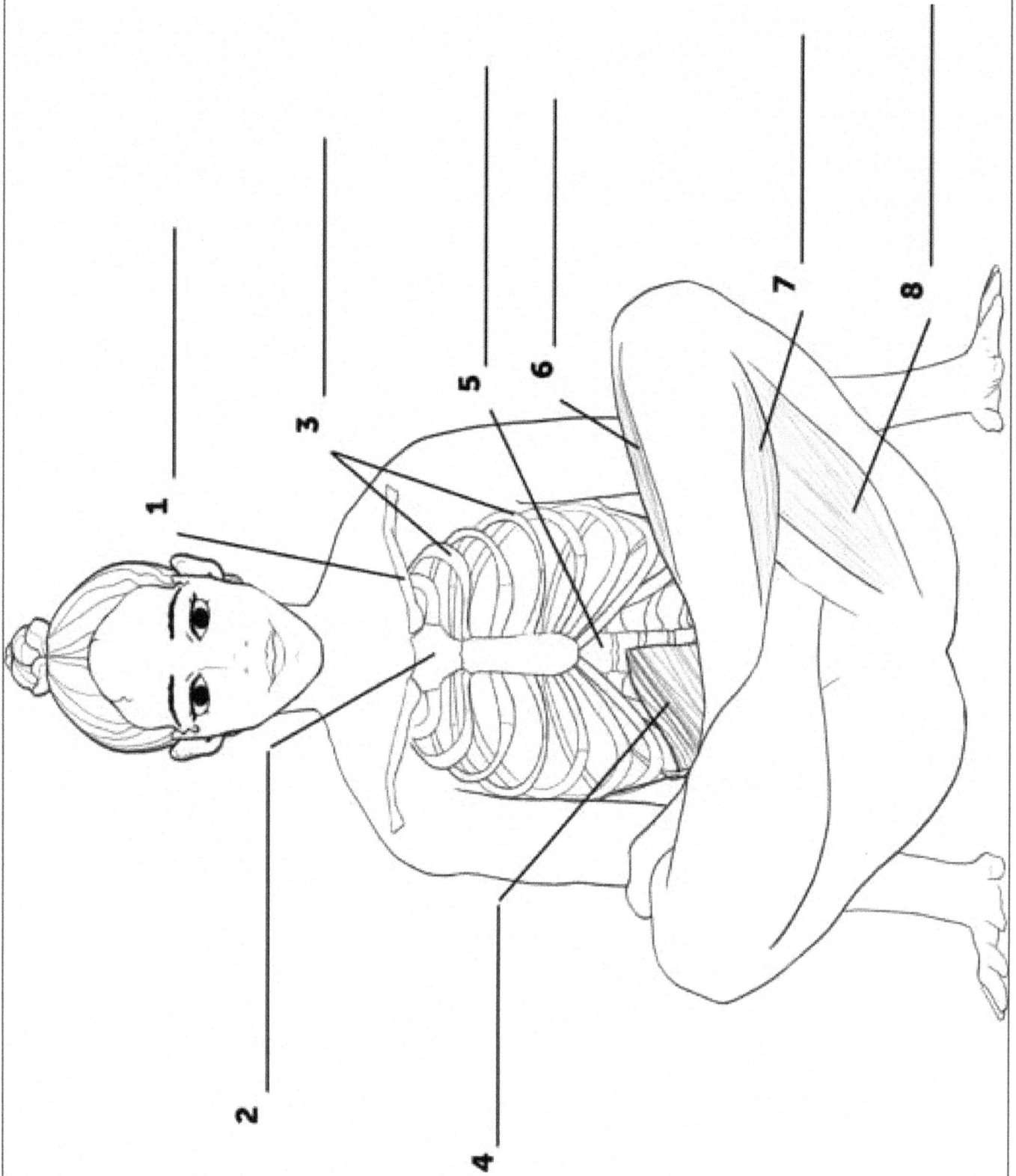

90. SCALE POSE

90. SCALE POSE

1. COLLARBONE
2. STERNUM
3. RIBS
4. INTERNAL OBLIQUE
5. SPINE
6. GASTROCNEMIUS
7. GASTROCNEMIUS
8. HAMSTRINGS

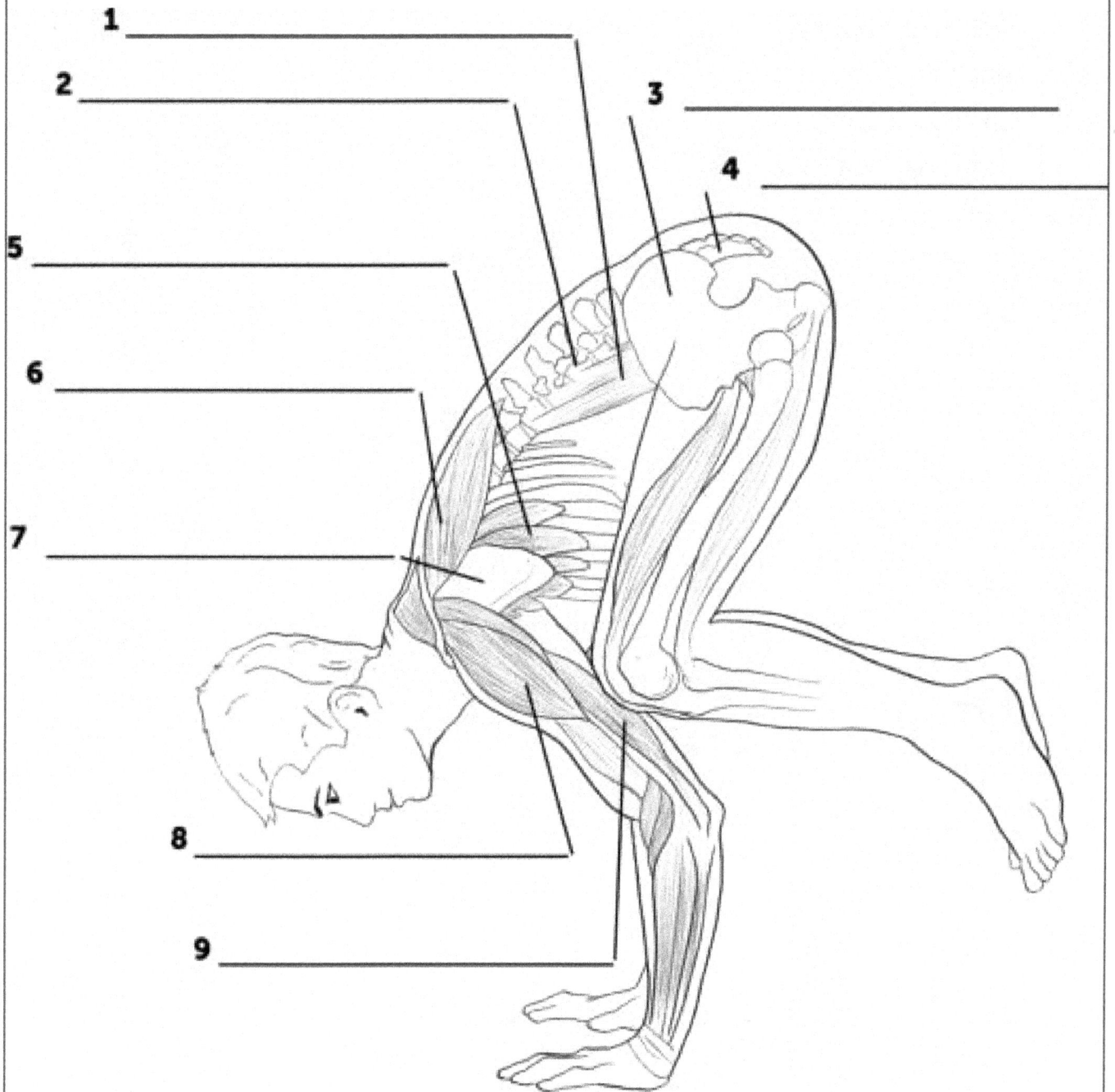

91. CROW POSE

1 _____

2 _____

3 _____

4 _____

5 _____

6 _____

7 _____

8 _____

9 _____

91. CROW POSE

1. PSOAS MAJOR
2. SPINE
3. PELVIS
4. SACRUM
5. SERRATUS ANTERIOR
6. TRAPEZIUS
7. SCAPULA
8. DELTOID
9. TRICEPS BRACHII

92. FOUR LIMBED STAFF POSE

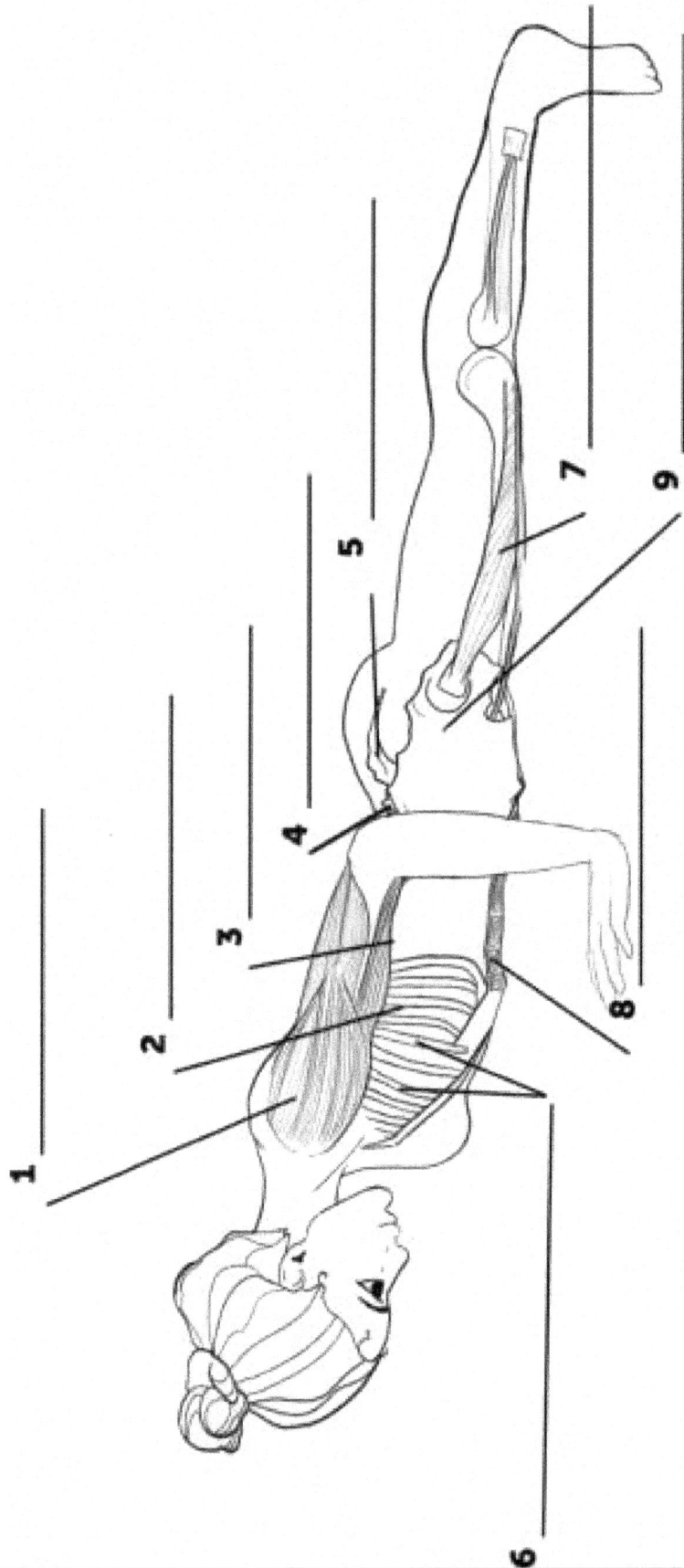

92. FOUR LIMBED STAFF POSE

1. DELTOID
2. RIBS
3. BICEPS BRACHII
4. SPINE
5. SACRUM
6. RIBS
7. RECTUS FEMORIS
8. RECTUS ABDOMINIS
9. PELVIS

93. SIDE CROW POSE

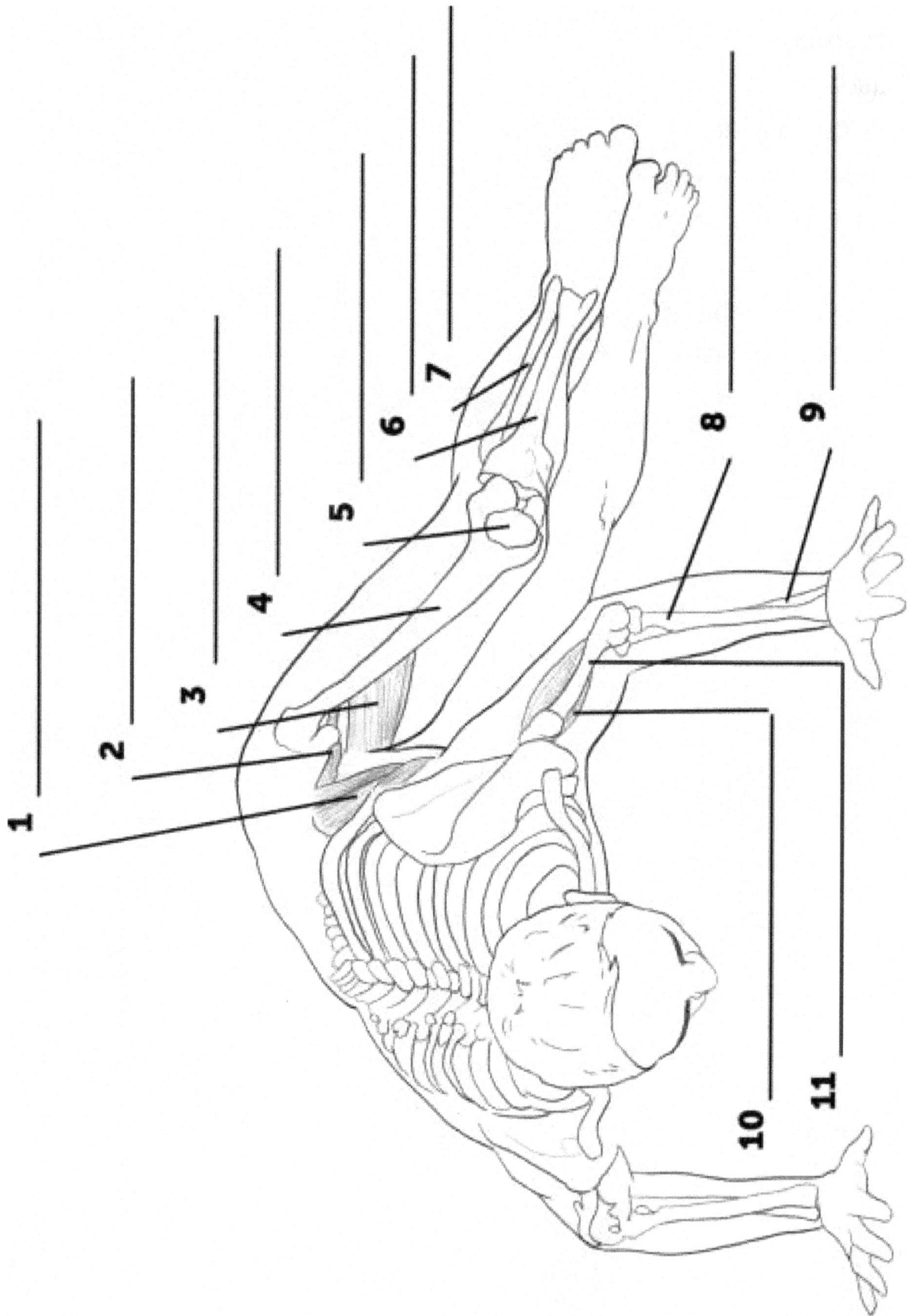

1

2

3

4

5

6

7

8

9

10

11

93. SIDE CROW POSE

1. EXTERNAL OBLIQUE
2. PECTINEUS
3. ADDUCTOR BREVIS
4. FEMUR
5. PATELLA
6. TIBIA
7. FIBULA
8. RADIUS
9. ULNA
10. TRICEPS BRACHII
11. HUMERUS

94. HALF BOAT POSE

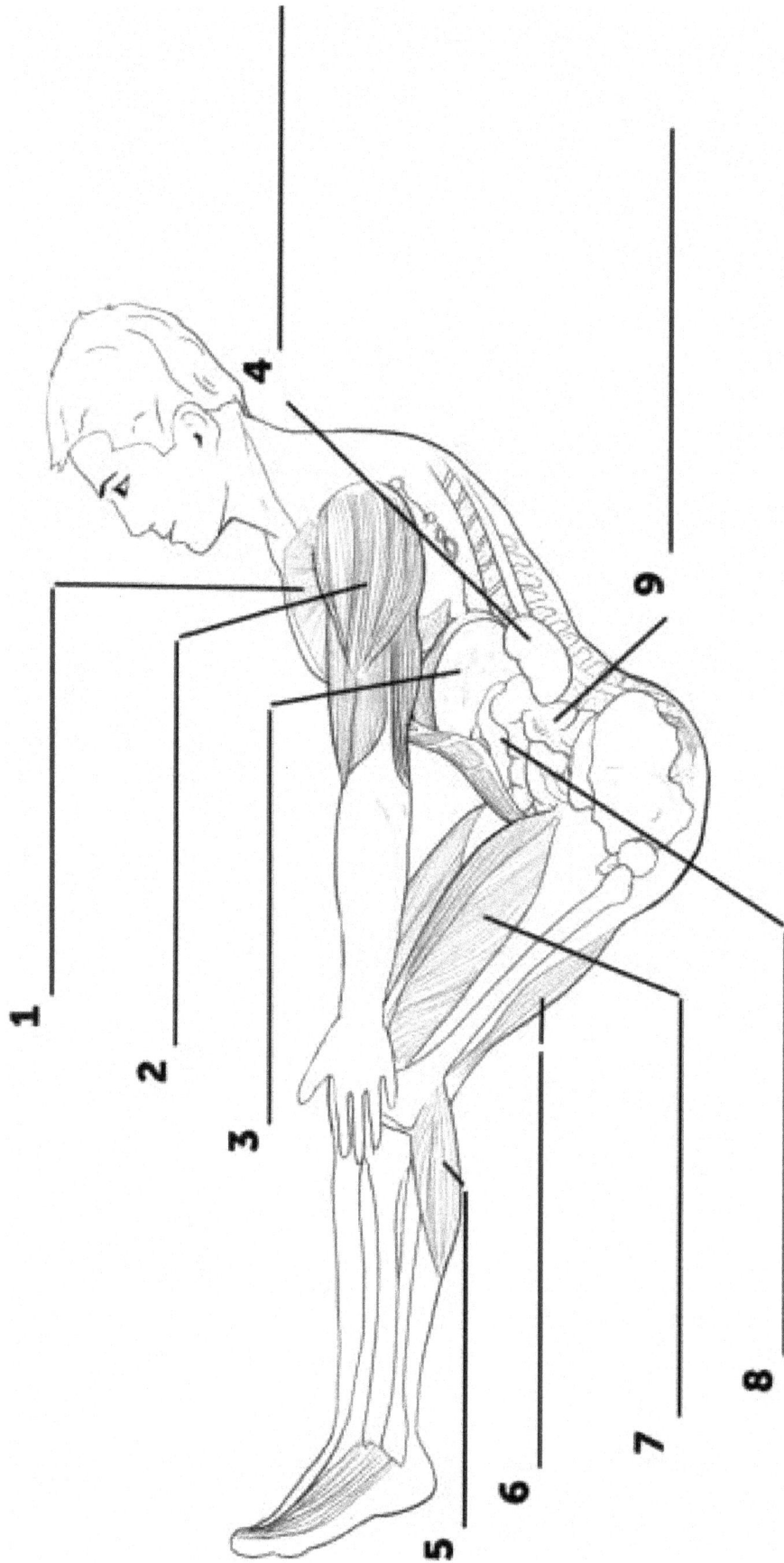

94. HALF BOAT POSE

1. PECTORALIS MAJOR
2. DELTOID
3. LIVER
4. KIDNEY
5. GASTROCNEMIUS
6. HAMSTRINGS
7. QUADRICEPS
8. STOMACH
9. ASCENDING COLON

95. FULL BOAT POSE

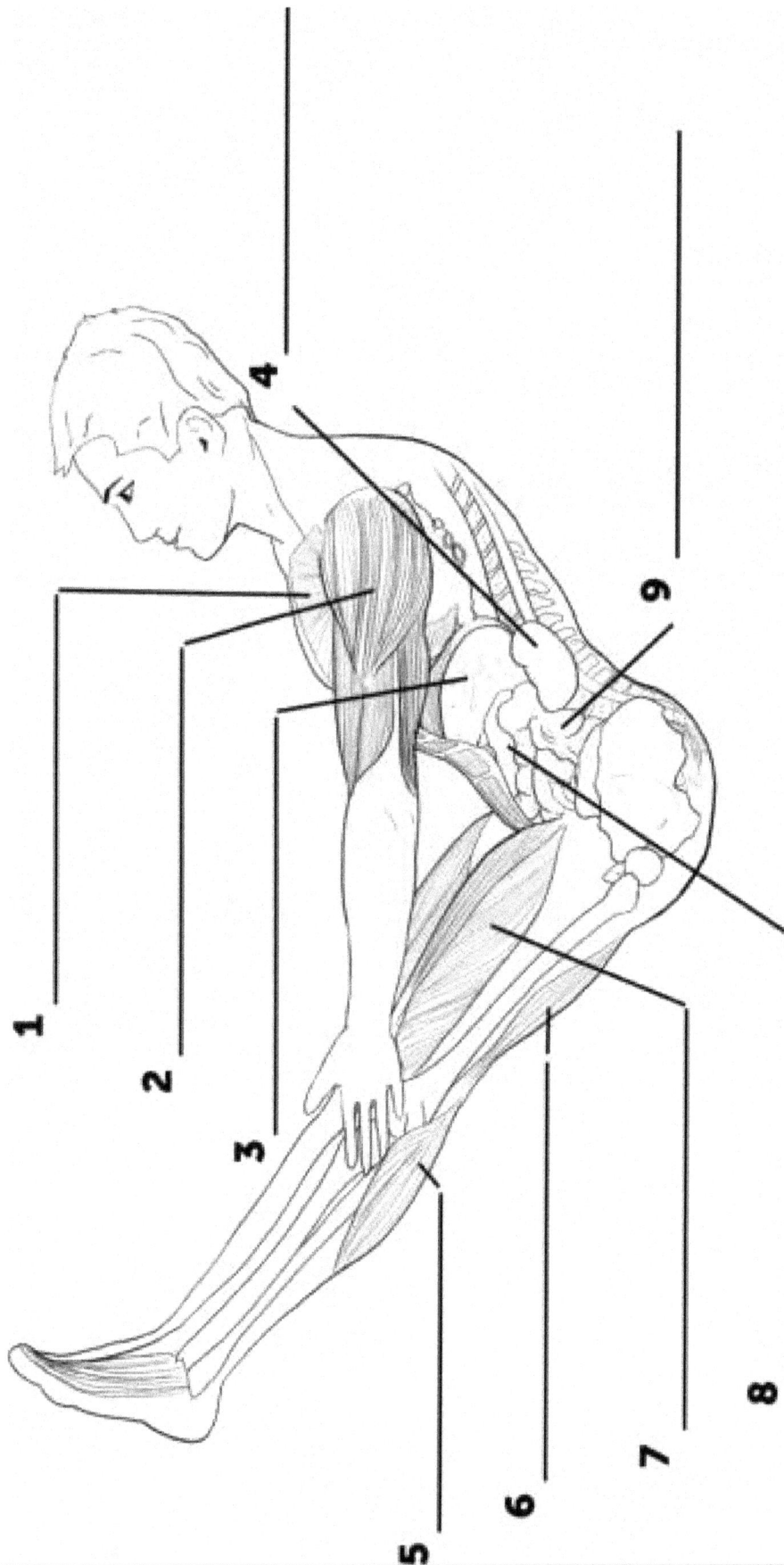

1

2

3

4

5

6

7

8

9

95. FULL BOAT POSE

1. PECTORALIS MAJOR
2. DELTOID
3. LIVER
4. KIDNEY
5. GASTROCNEMIUS
6. HAMSTRINGS
7. QUADRICEPS
8. STOMACH
9. ASCENDING COLON

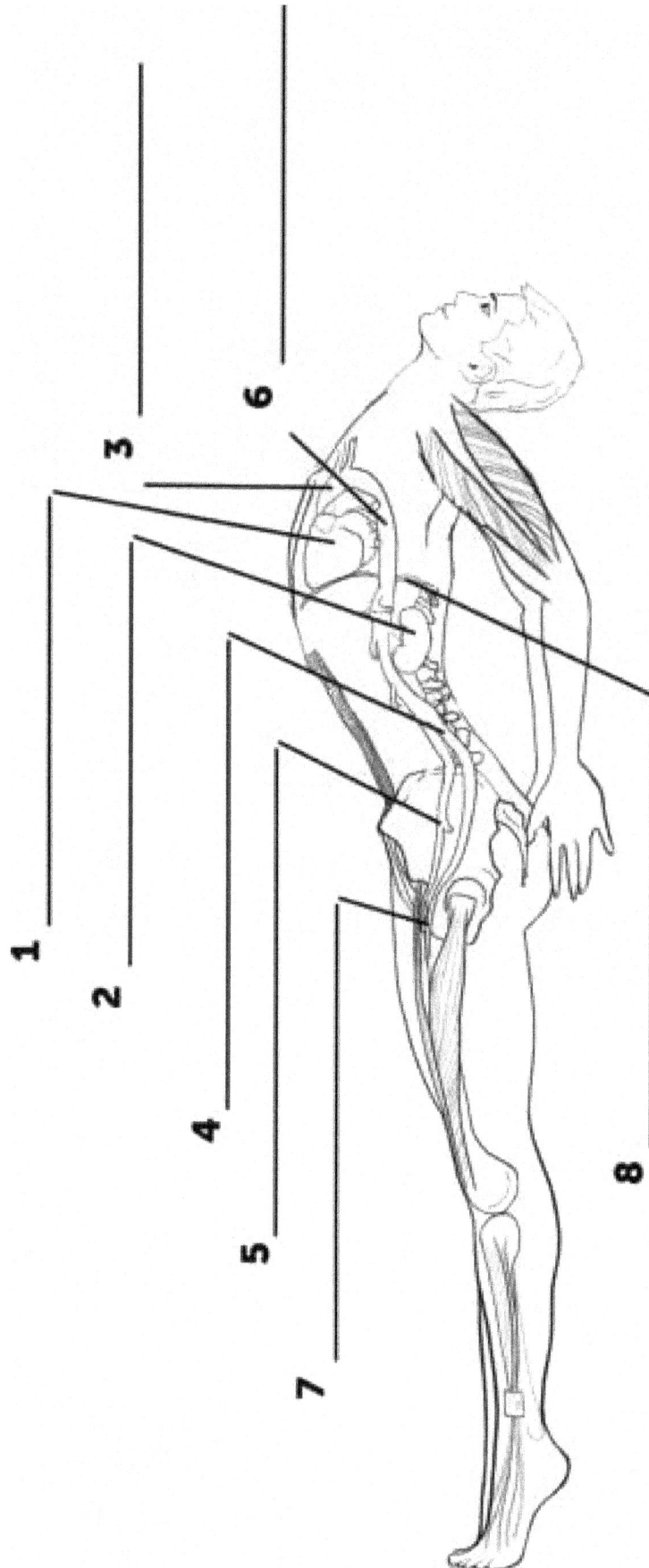

96. FISH POSE

1

2

3

4

5

6

7

8

96. FISH POSE

1. HEART
2. KIDNEY
3. ASCENDING THORACIC AORTA
4. ABDOMINAL AORTA
5. COMMON ILIAC ARTERY
6. DESCENDING THORACIC AORTA
7. FEMORAL ARTERY
8. DIAPHRAGM

97. SUPPORTED HEADSTAND POSE

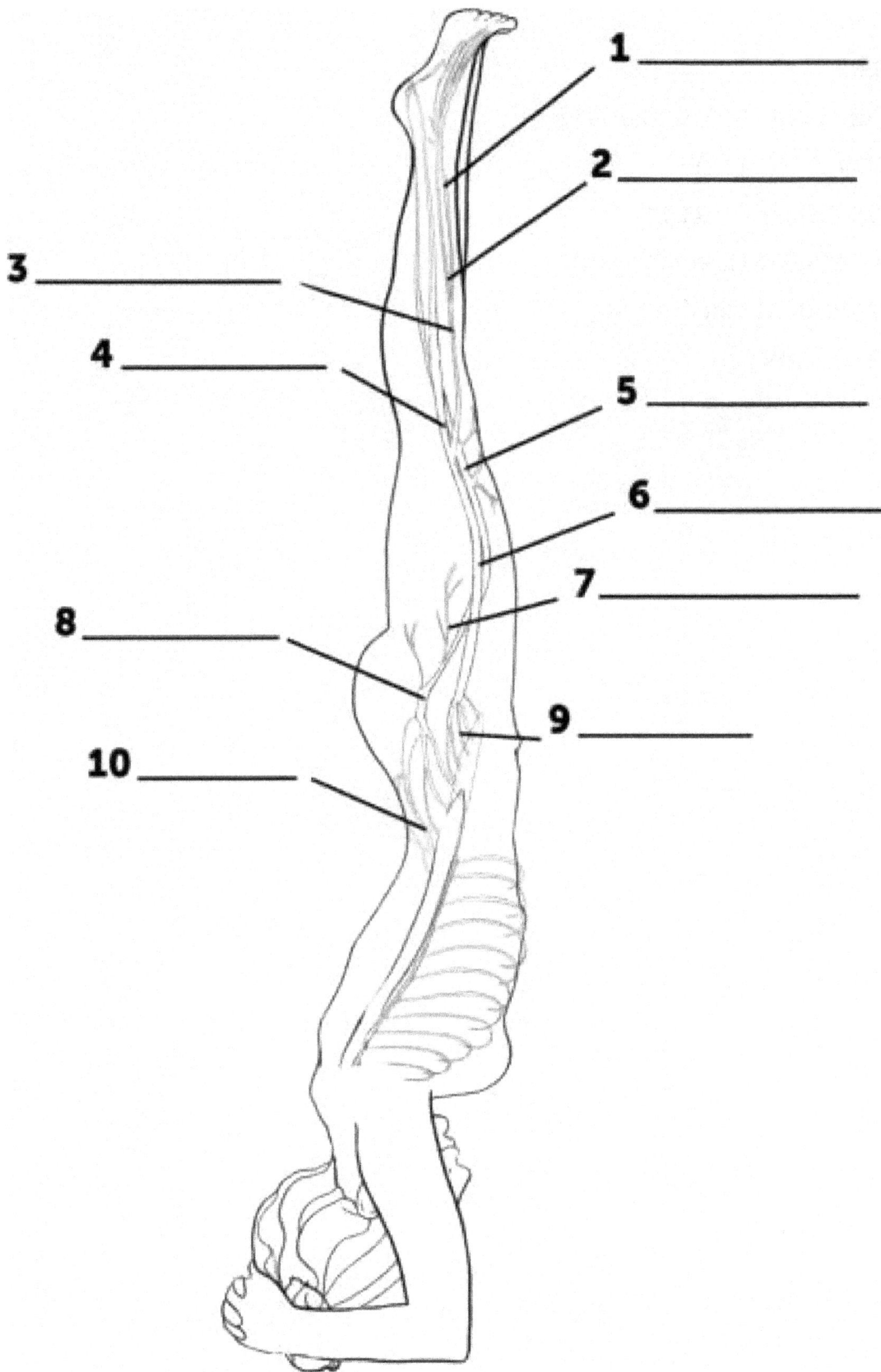

1 _____

2 _____

3 _____

4 _____

5 _____

6 _____

7 _____

8 _____

9 _____

10 _____

97. SUPPORTED HEADSTAND POSE

1.   SUPERFICIAL PERONEAL

2.   DEEP PERONEAL

3.   COMMON PERONEAL

4.   TIBIAL

5.   SAPHENOUS

6.   SCIATIC

7.   MUSCULAR BRANCHES OF FEMORAL

8.   FEMORAL

9.   SACRAL PLEXUS

10. LUMBAR PLEXUS

98. SUPPORTED SHOULDER STAND

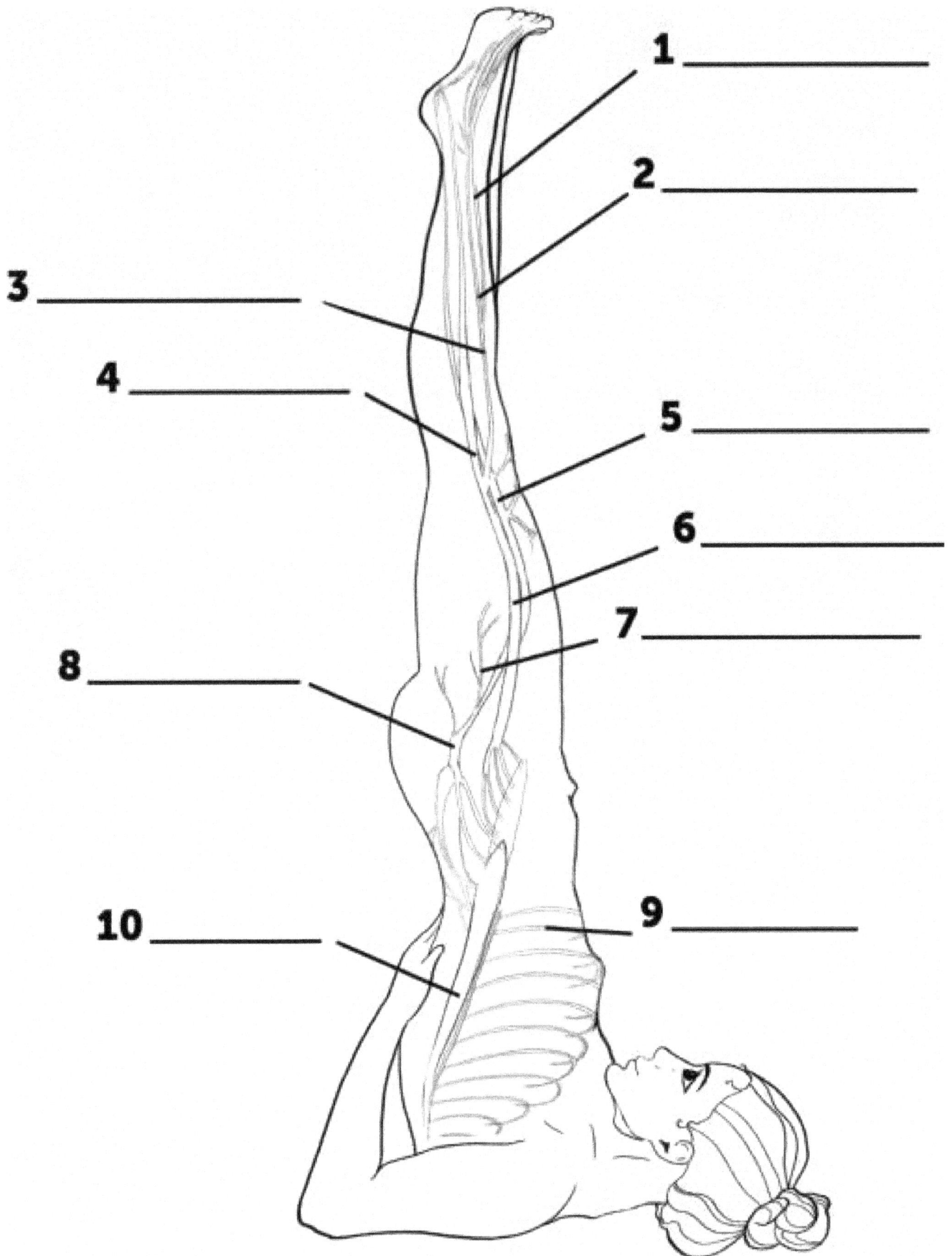

1 _____

2 _____

3 _____

4 _____

5 _____

6 _____

7 _____

8 _____

9 _____

10 _____

98. SUPPORTED SHOULDER STAND

1. SUPERFICIAL PERONEAL

2. DEEP PERONEAL

3. COMMON PERONEAL

4. TIBIAL

5. SAPHENOUS

6. SCIATIC

7. MUSCULAR BRANCHES OF FEMORAL

8. FEMORAL

9. INTERCOSTALS

10. SPINAL CORD

99. PLOW POSE

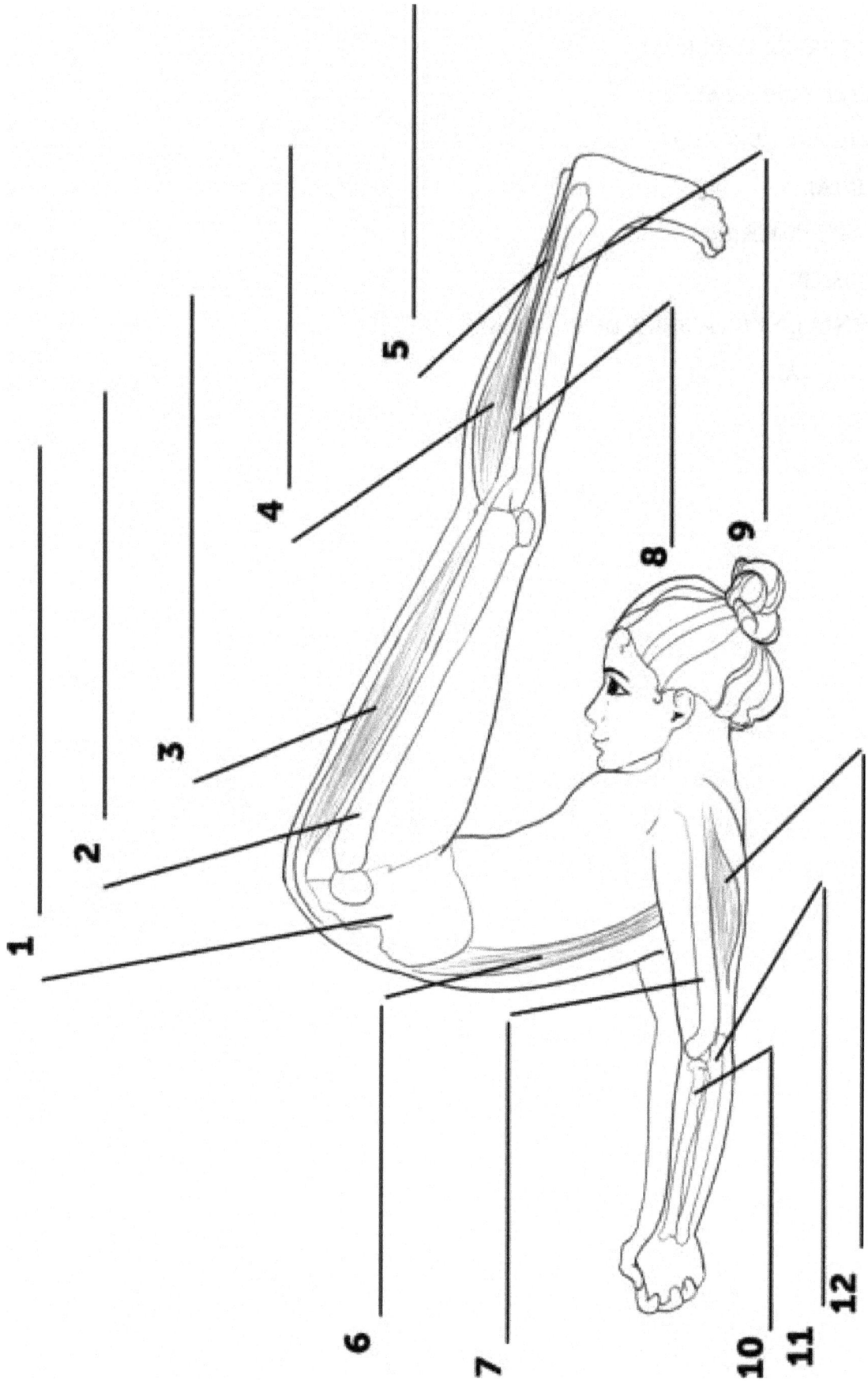

99. PLOW POSE

1. PELVIS

2. FEMUR

3. HAMSTRINGS

4. GASTROCNEMIUS

5. SOLEUS

6. ERECTOR SPINAE

7. HUMERUS

8. FIBULA

9. TIBIA

10. RADIUS

11. ULNAS

12. TRICEPS BRACHII

100. KNEE-TO-EAR POSE

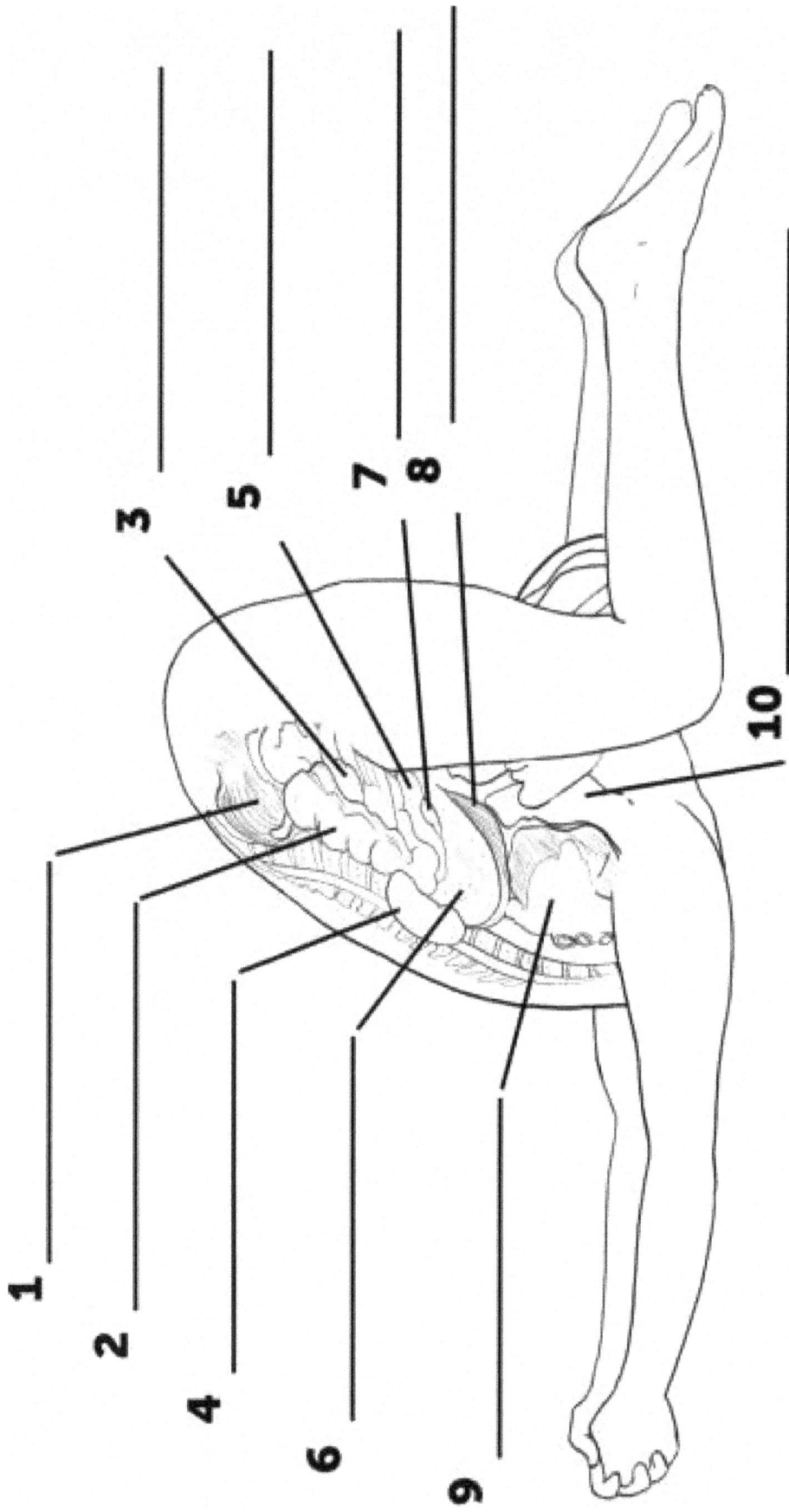

100. KNEE-TO-EAR POSE

1. RECTUM
2. ASCENDING COLON
3. COILS OF SMALL INTESTINE
4. KIDNEY
5. STOMACH
6. LIVER
7. GALLBLADDER
8. DIAPHRAGM
9. HEART
10. LUNGS

101. HALF-MOON POSE

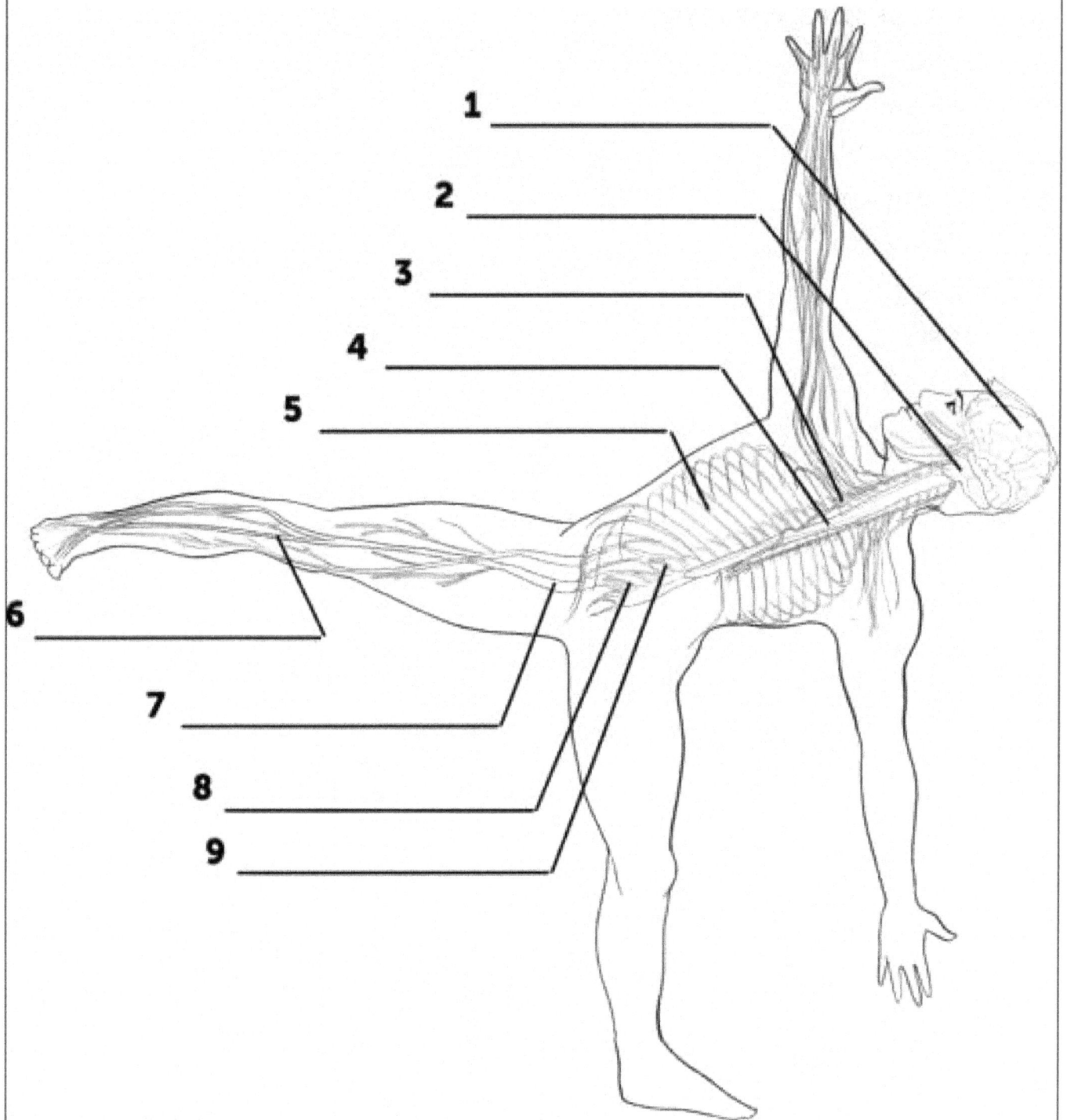

1 _____

2 _____

3 _____

4 _____

5 _____

6 _____

7 _____

8 _____

9 _____

101. HALF-MOON POSE

1. CEREBRUM
2. BRAINSTEM
3. BRACHIAL PLEXUS
4. SPINAL CORD
5. INTERCOSTALS
6. TIBIAL
7. SCIATIC
8. SACRAL PLEXUS
9. LUMBAR PLEXUS

102. COMPASS POSE

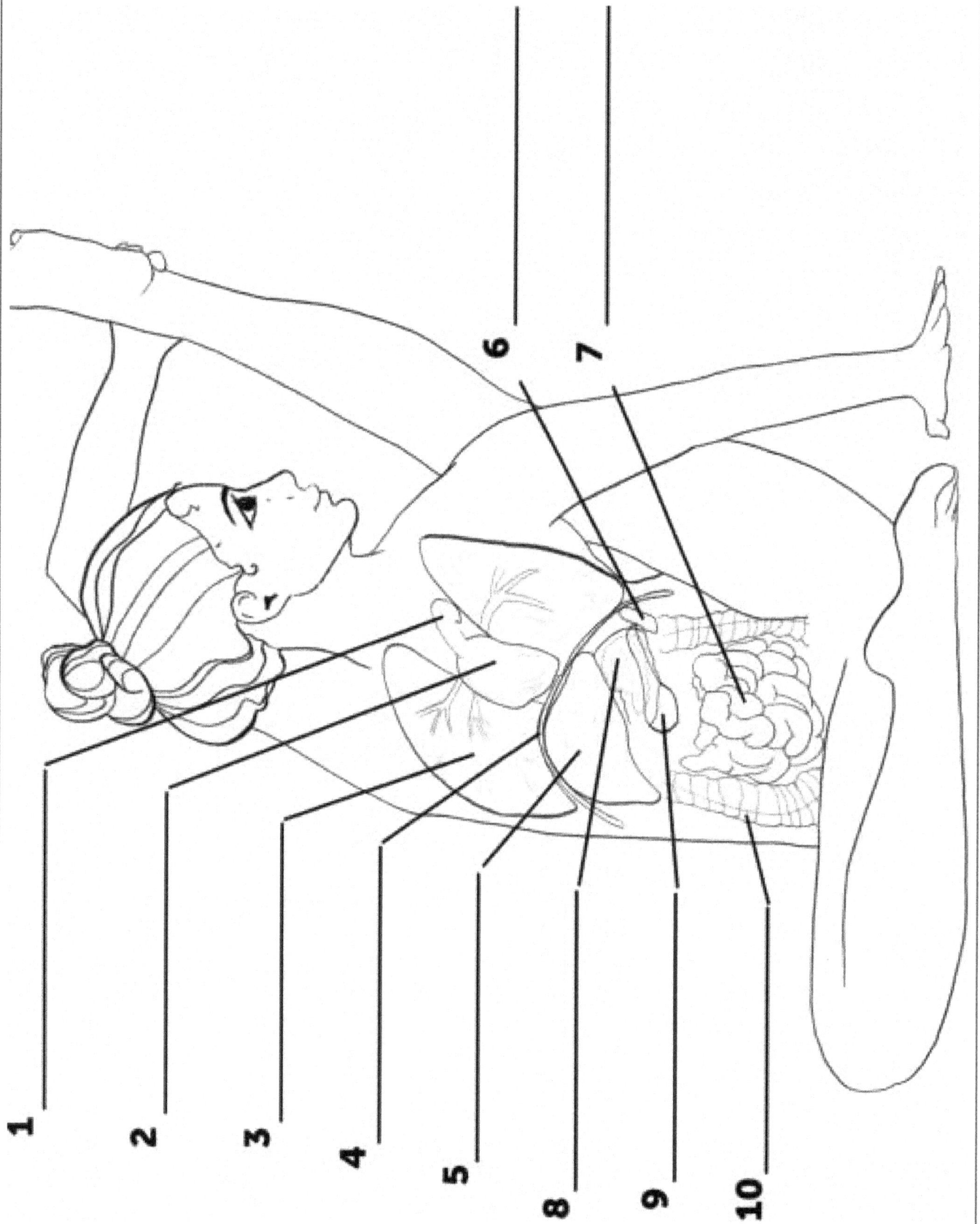

1

2

3

4

5

6

7

8

9

10

102. COMPASS POSE

1.   AORTA
2.   HEART
3.   LUNGS
4.   DIAPHRAGM
5.   LIVER
6.   SPLEEN
7.   COILS OF SMALL INTESTINE
8.   STOMACH
9.   PANCREAS
10. ASCENDING COLON

103. TWISTED HEAD-TO-KNEE POSE

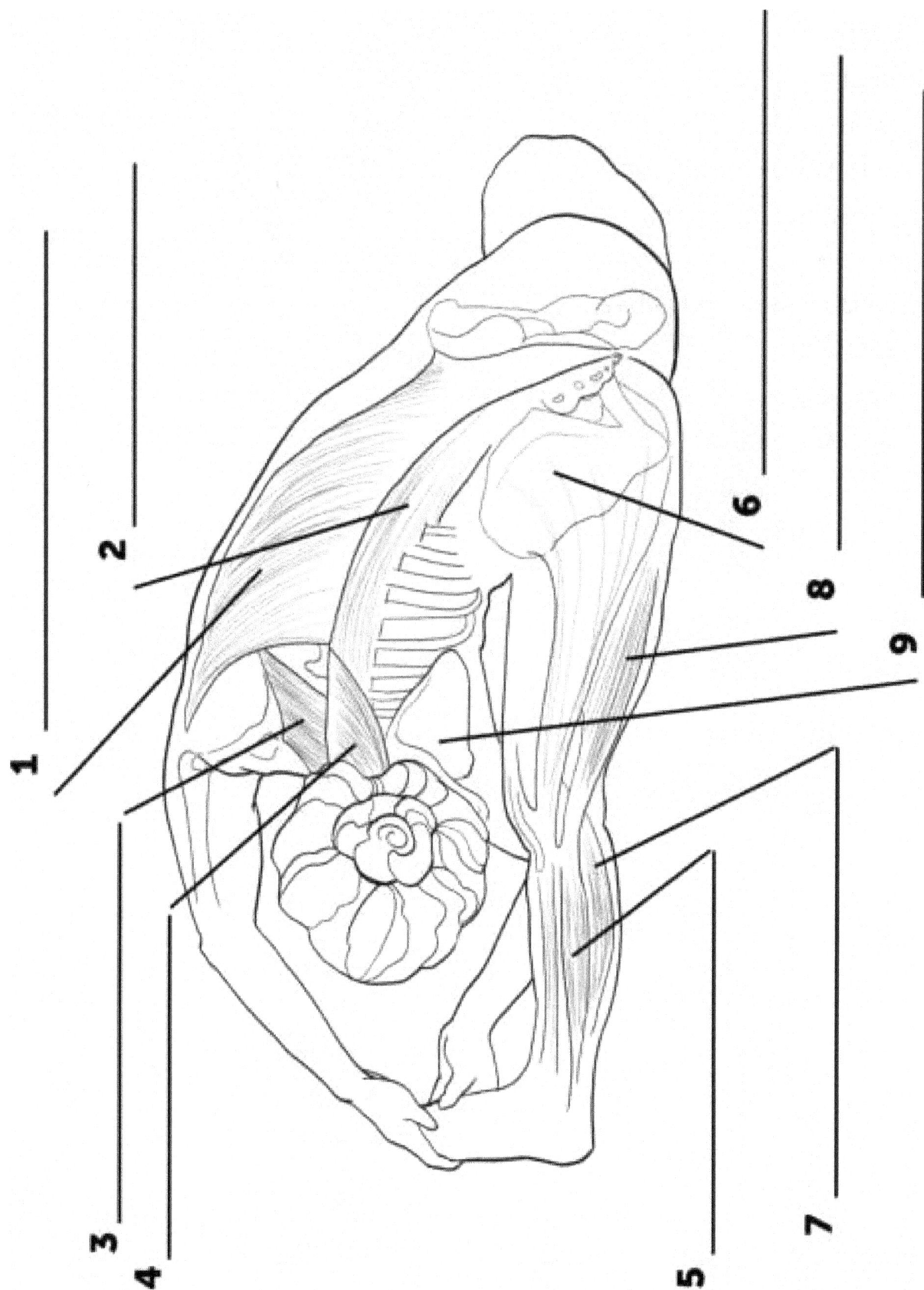

103. TWISTED HEAD-TO-KNEE POSE

1. LATISSIMUS DORSI
2. ERECTOR SPINAE
3. RHOMBOIDS
4. TRAPEZIUS
5. SOLEUS
6. PELVIS
7. GASTROCNEMIUS
8. HAMSTRINGS
9. SCAPULA

104. STANDING SPLIT POSE

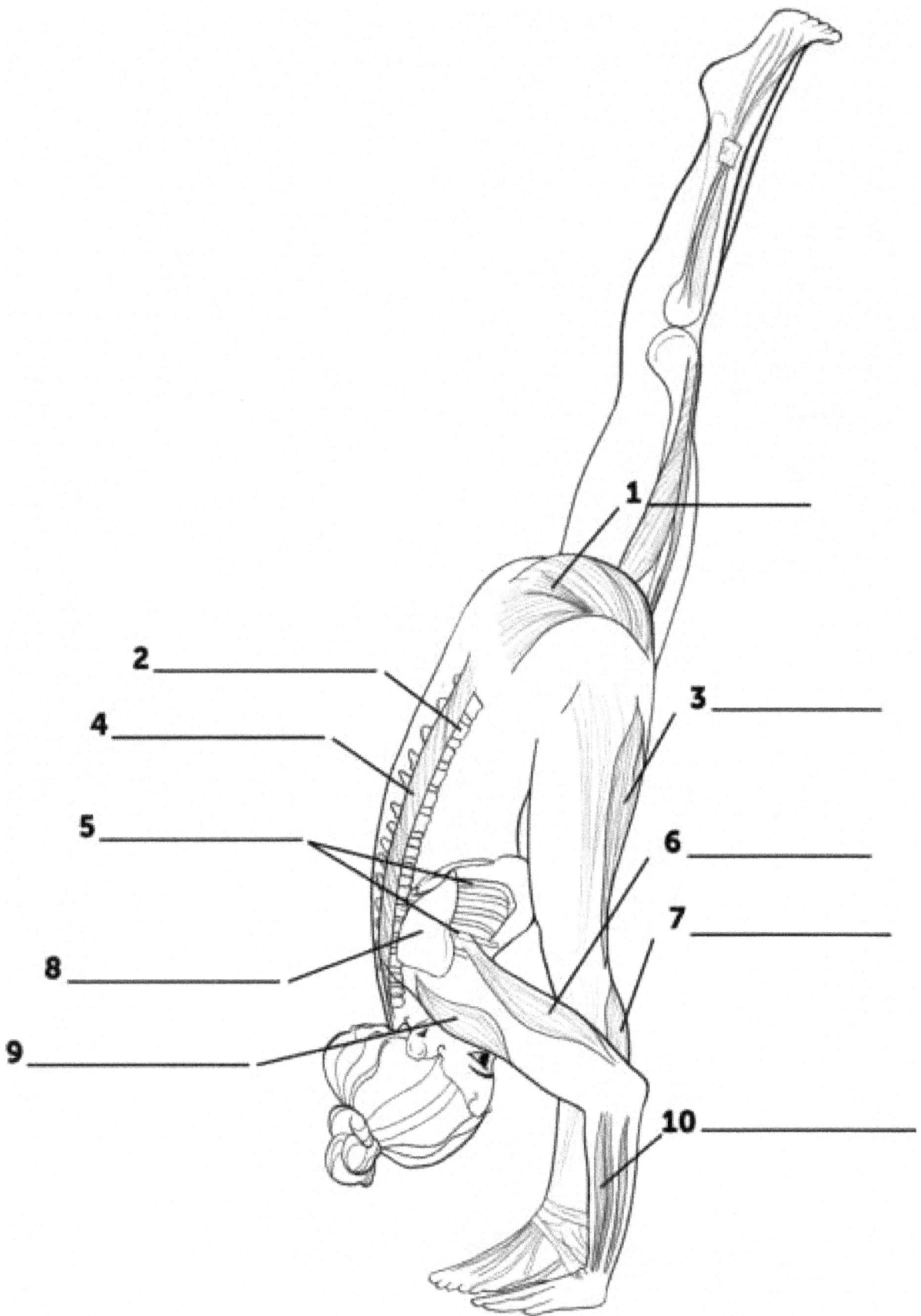

1 _____

2 _____

3 _____

4 _____

5 _____

6 _____

7 _____

8 _____

9 _____

10 _____

104. STANDING SPLIT POSE

1. PIRIFORMIS
2. SPINE
3. HAMSTRINGS
4. ERECTOR SPINAE
5. RIBS
6. TRICEPS BRACHII
7. GASTROCNEMIUS
8. SCAPULA
9. DELTOID
10. PRONATORS

105. ARCHER POSE

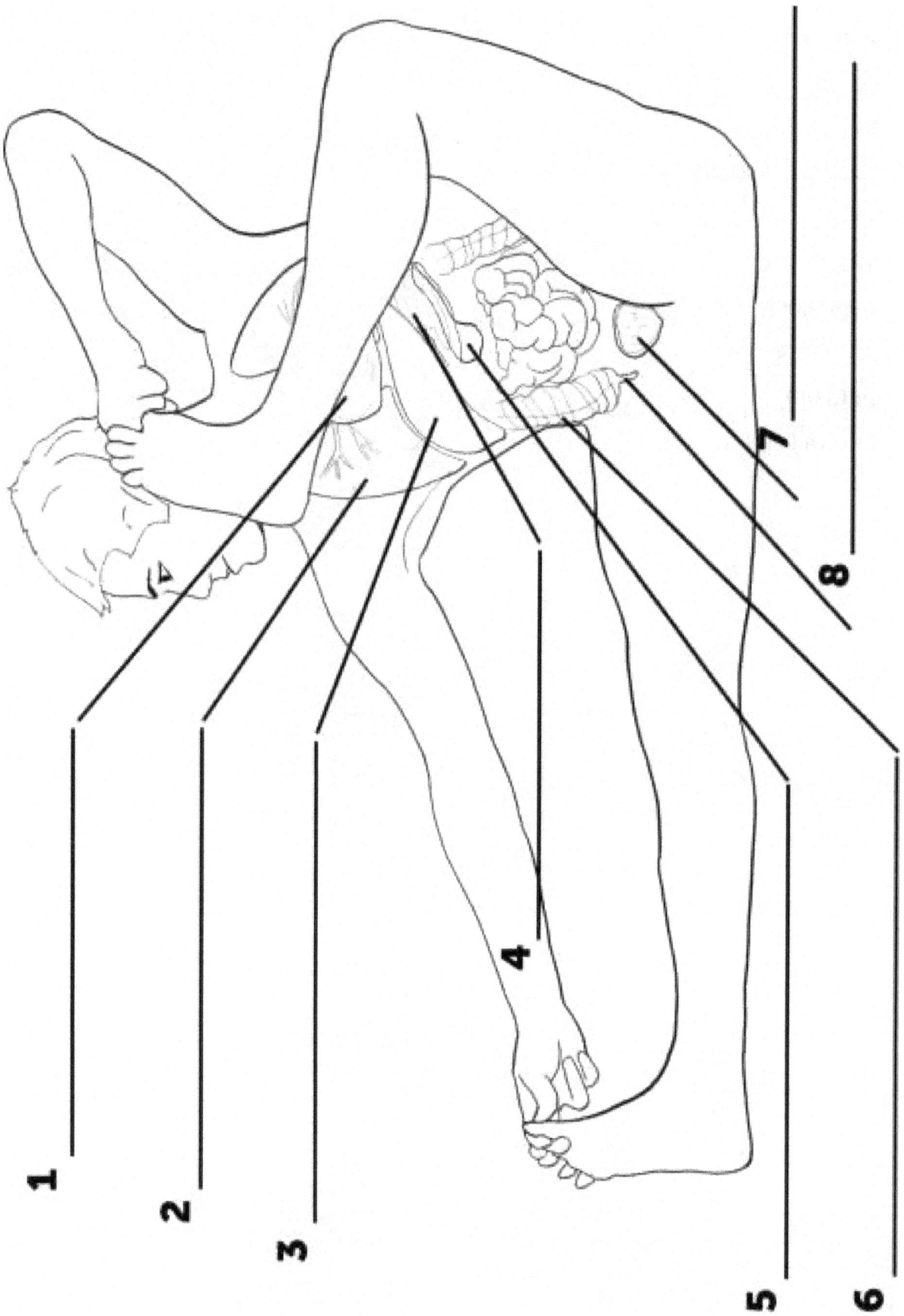

1

2

3

4

5

6

7

8

105. ARCHER POSE

1. HEART
2. LUNGS
3. LIVER
4. STOMACH
5. PANCREAS
6. ASCENDING COLON
7. URINARY BLADDER
8. APPENDIX

106. YOGA HANDSTAND POSE

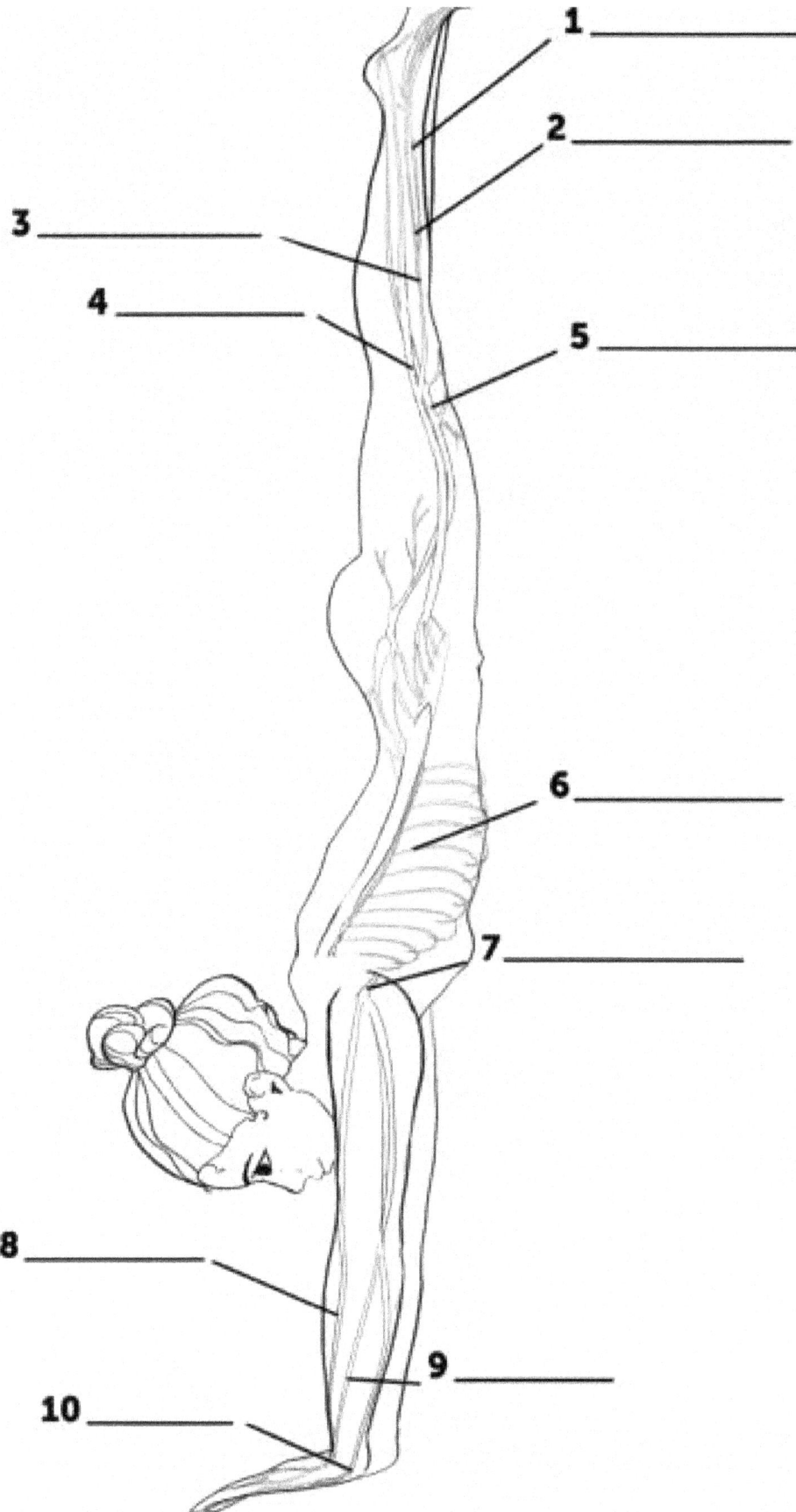

1 _____

2 _____

3 _____

4 _____

5 _____

6 _____

7 _____

8 _____

9 _____

10 _____

106. YOGA HANDSTAND POSE

1. SUPERFICIAL PERONEAL

2. DEEP PERONEAL

3. COMMON PERONEAL

4. TIBIAL

5. SAPHENOUS

6. INTERCOSTALS

7. BRACHIAL PLEXUS

8. RADIAL

9. MEDIAN

10. ULNAR

107. ELEPHANT TRUNK POSE

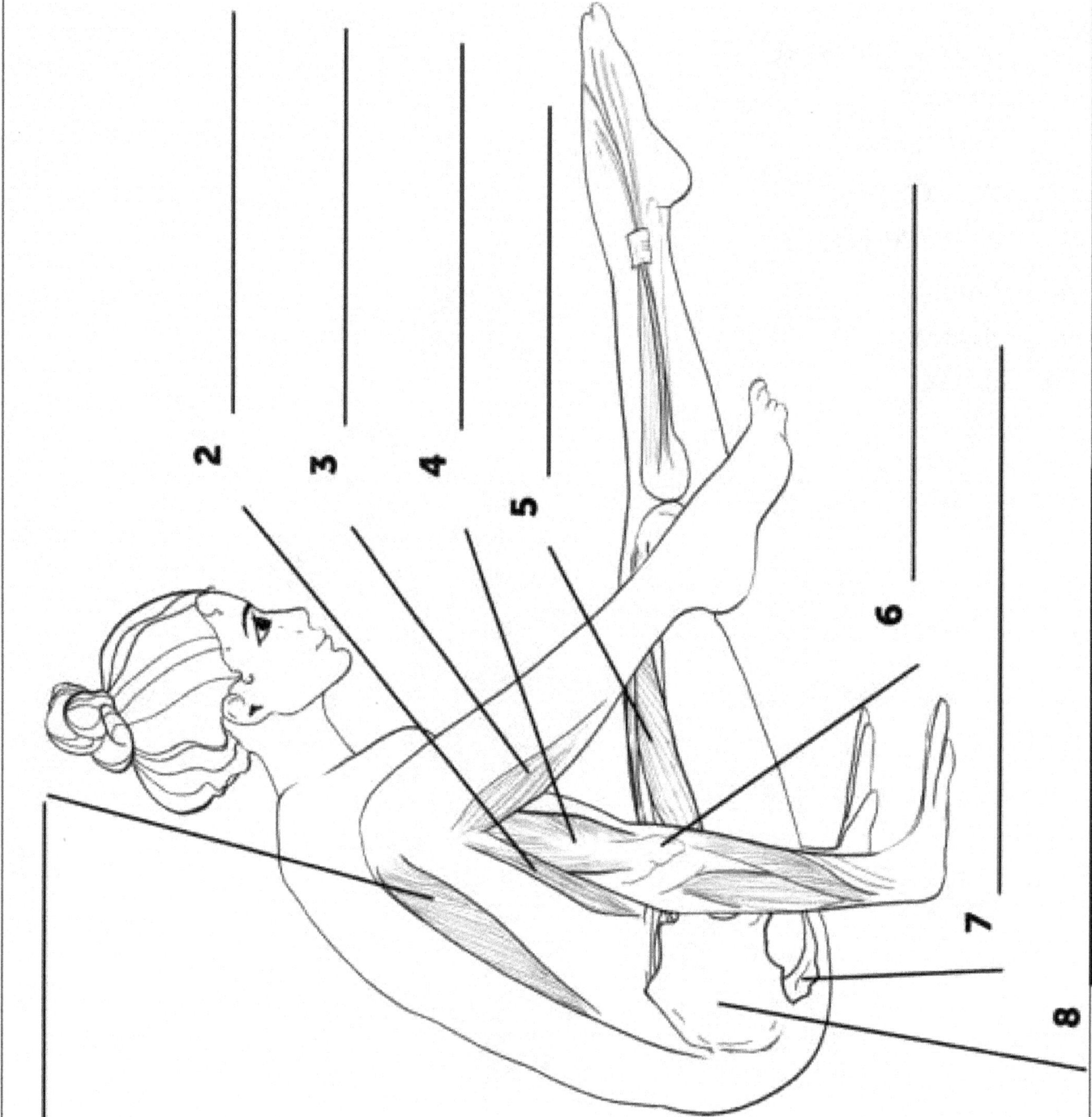

1

2

3

4

5

6

7

8

107. ELEPHANT TRUNK POSE

1. RECTUS FEMORIS
2. HAMSTRINGS
3. GASTROCNEMIUS
4. TRICEPS BRACHII
5. QUADRICEPS
6. ELBOW
7. SACRUM
8. PELVIS

YOGA POSES FOR EXPERTS

108. FULL LORD OF THE FISHES POSE

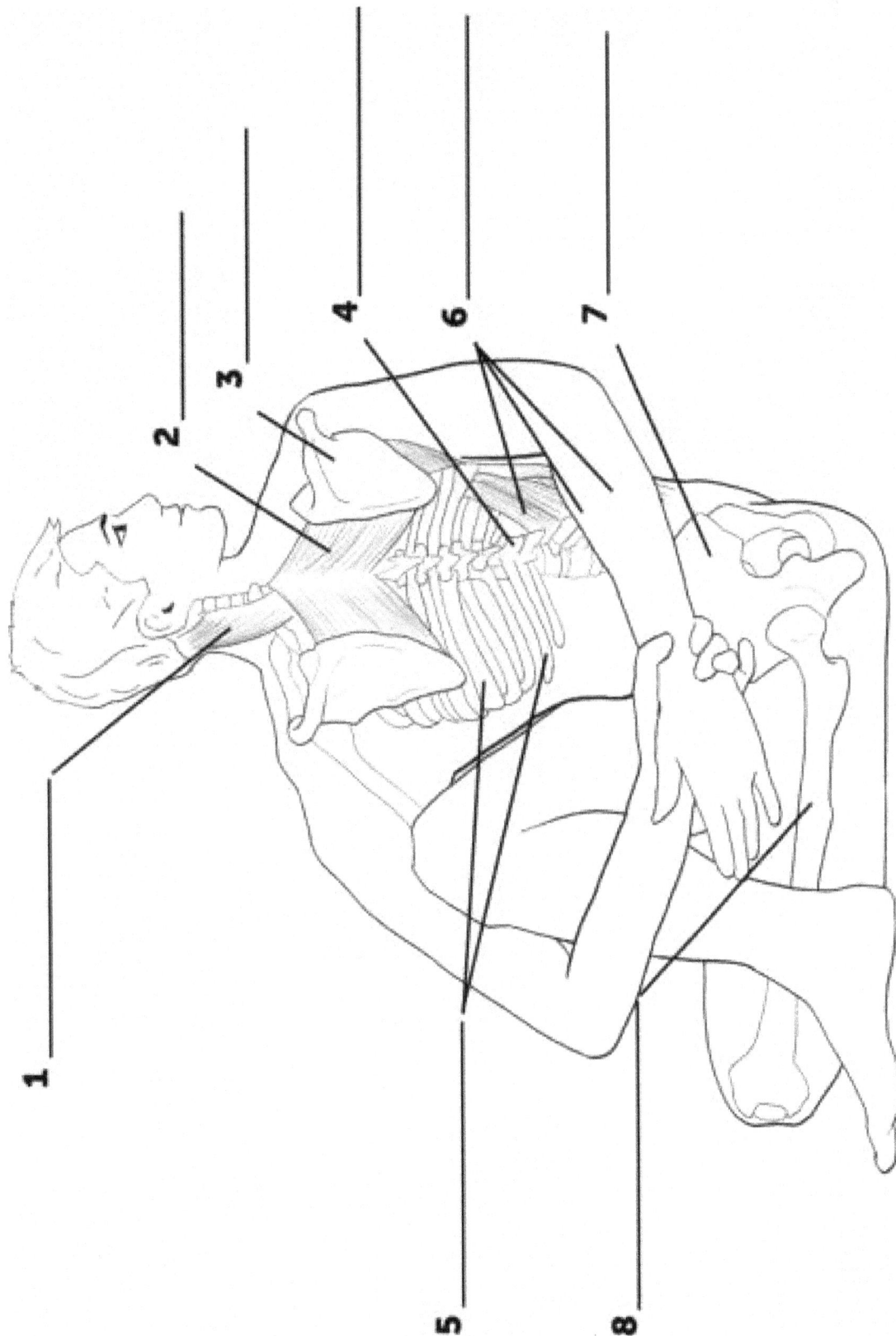

1
2
3
4
5
6
7
8

108. FULL LORD OF THE FISHES POSE

1. SPLENIUS CAPITIS
2. RHOMBOIDS
3. SCAPULA
4. SPINE
5. RIBS
6. ERECTOR SPINAE
7. PELVIS
8. FEMUR

109. FLYING CROW POSE

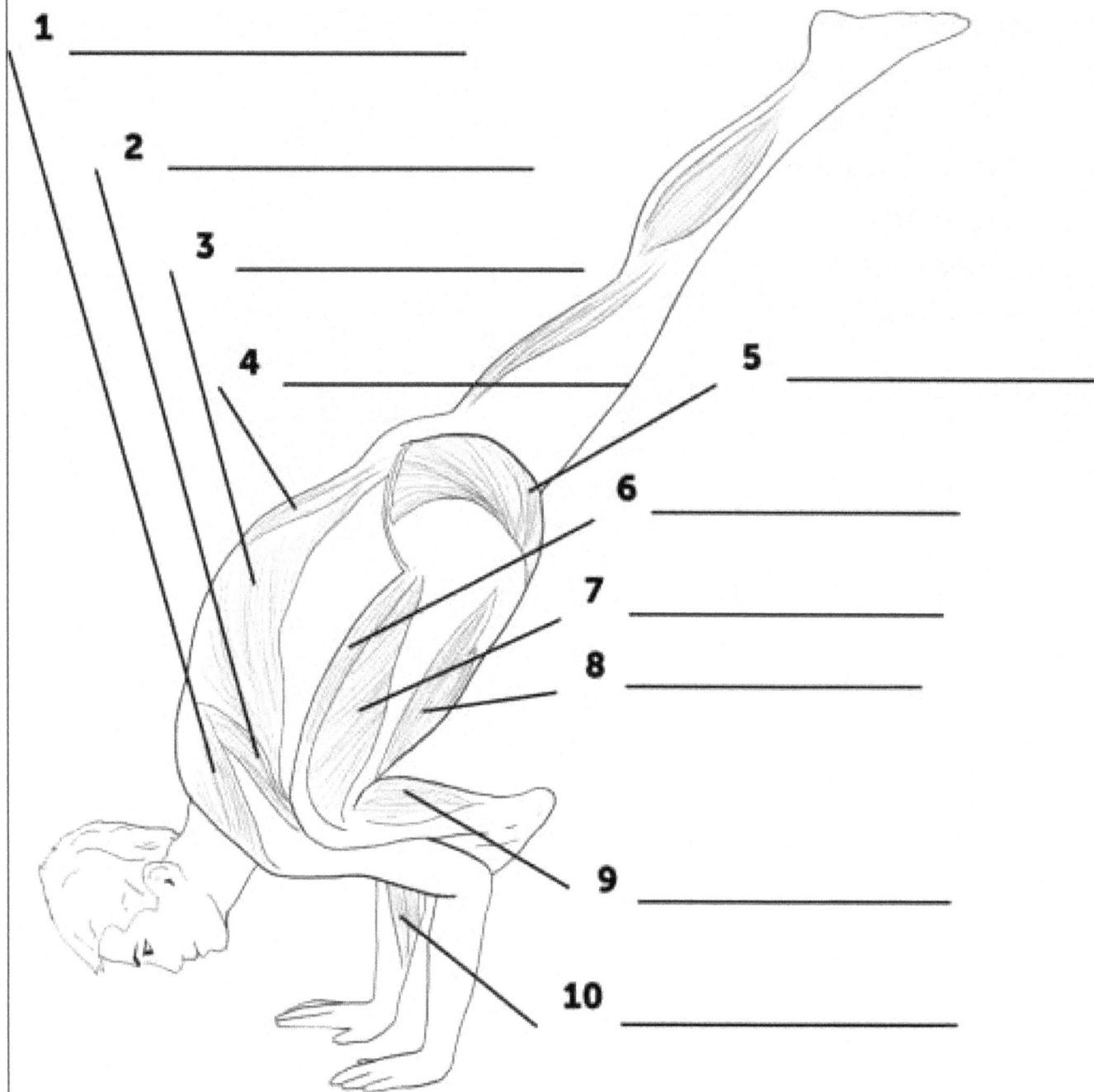

1 _____

2 _____

3 _____

4 _____

5 _____

6 _____

7 _____

8 _____

9 _____

10 _____

109. FLYING CROW POSE

1. DELTOID
2. TRICEPS BRACHII
3. LATISSIMUS DORSI
4. ERECTOR SPINAE
5. GLUTEUS MAXIMUS
6. RECTUS FEMORIS
7. VASTUS LATERALIS
8. HAMSTRINGS
9. GASTROCNEMIUS
10. PRONATORS

110. SCORPION POSE

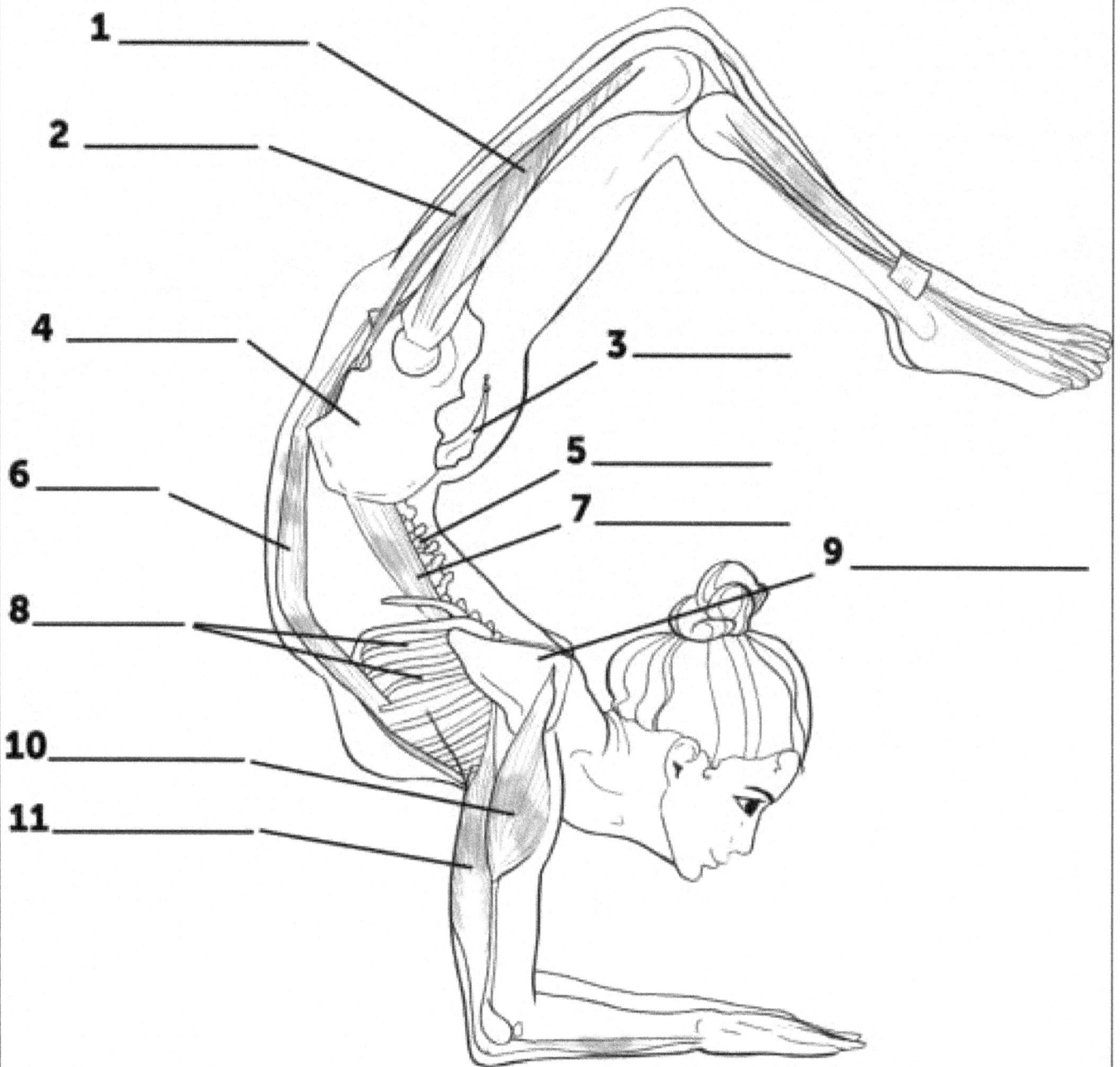

1 _____

2 _____

4 _____

3 _____

6 _____

5 _____

7 _____

9 _____

8 _____

10 _____

11 _____

110. SCORPION POSE

1.   VASTUS LATERALIS
2.   RECTUS FEMORIS
3.   SACRUM BONE
4.   PELVIS
5.   SPINE
6.   RECTUS ABDOMINIS
7.   PSOAS MAJOR
8.   RIBS
9.   SCAPULA
10. DELTOID
11. TRICEPS BRACHII

111. FIREFLY POSE

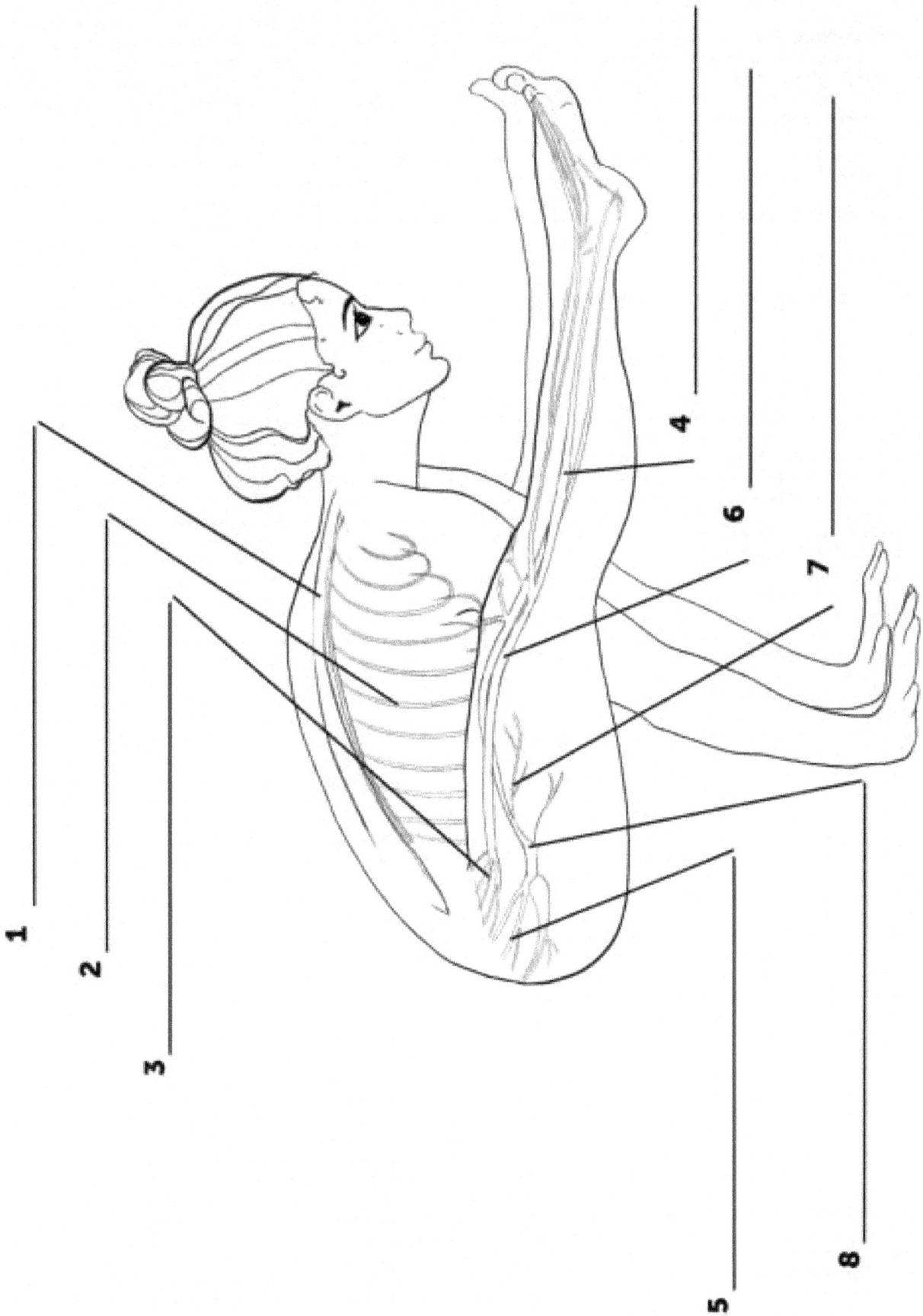

1
2
3
4
5
6
7
8

111. FIREFLY POSE

1. SPINAL CORD
2. INTERCOSTALS
3. SACRAL PLEXUS
4. TIBIAL
5. LUMBAR PLEXUS
6. SCIATIC
7. MUSCULAR BRANCHES OF FEMORAL
8. FEMORAL

112. BIRD OF PARADISE POSE

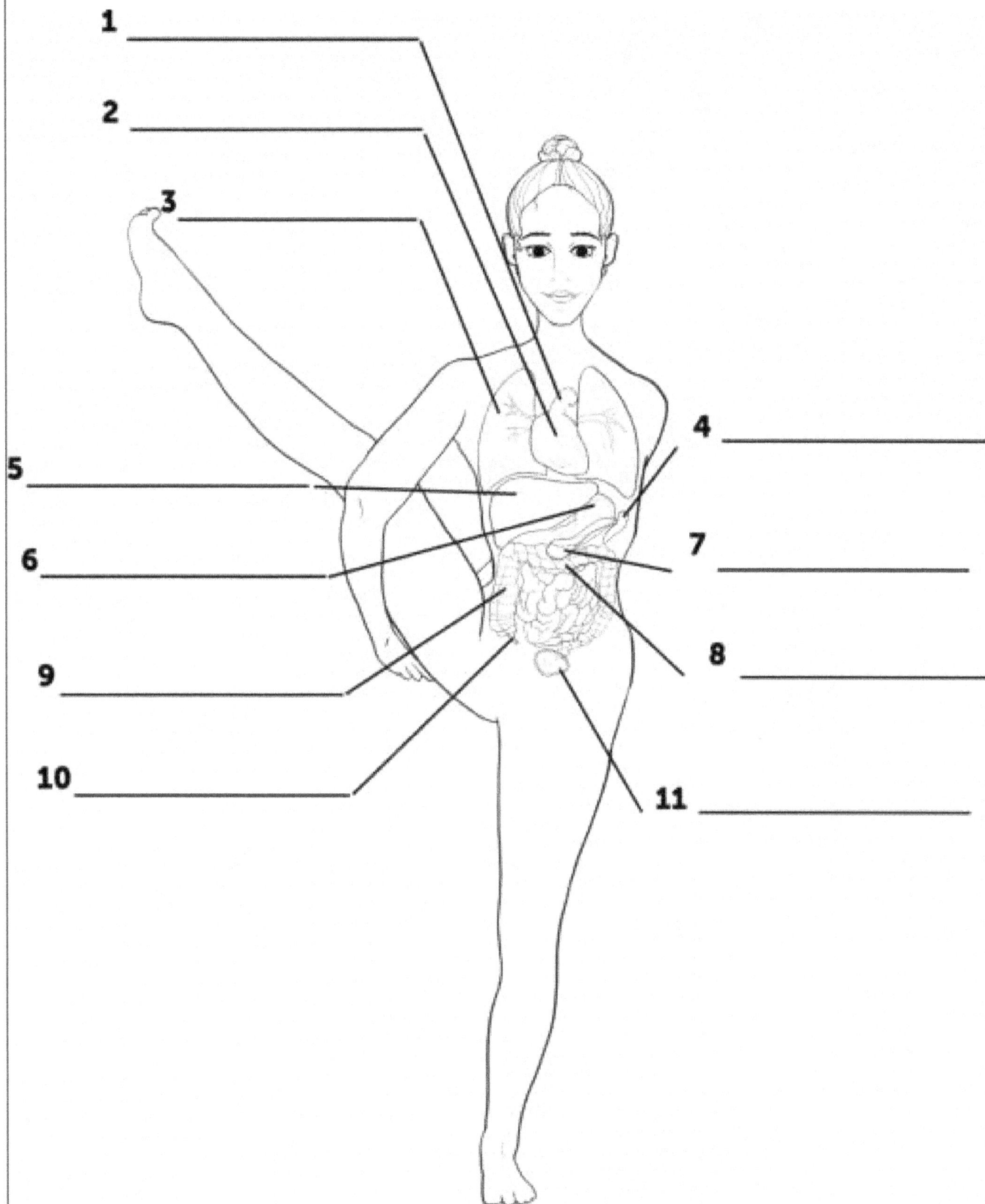

1 _____

2 _____

3 _____

4 _____

5 _____

6 _____

7 _____

8 _____

9 _____

10 _____

11 _____

112. BIRD OF PARADISE POSE

1. AORTA
2. HEART
3. LUNGS
4. SPLEEN
5. LIVER
6. STOMACH
7. PANCREAS
8. TRANSVERSE COLON
9. ASCENDING COLON
10. APPENDIX
11. URINARY BLADDER

113. PEACOCK POSE

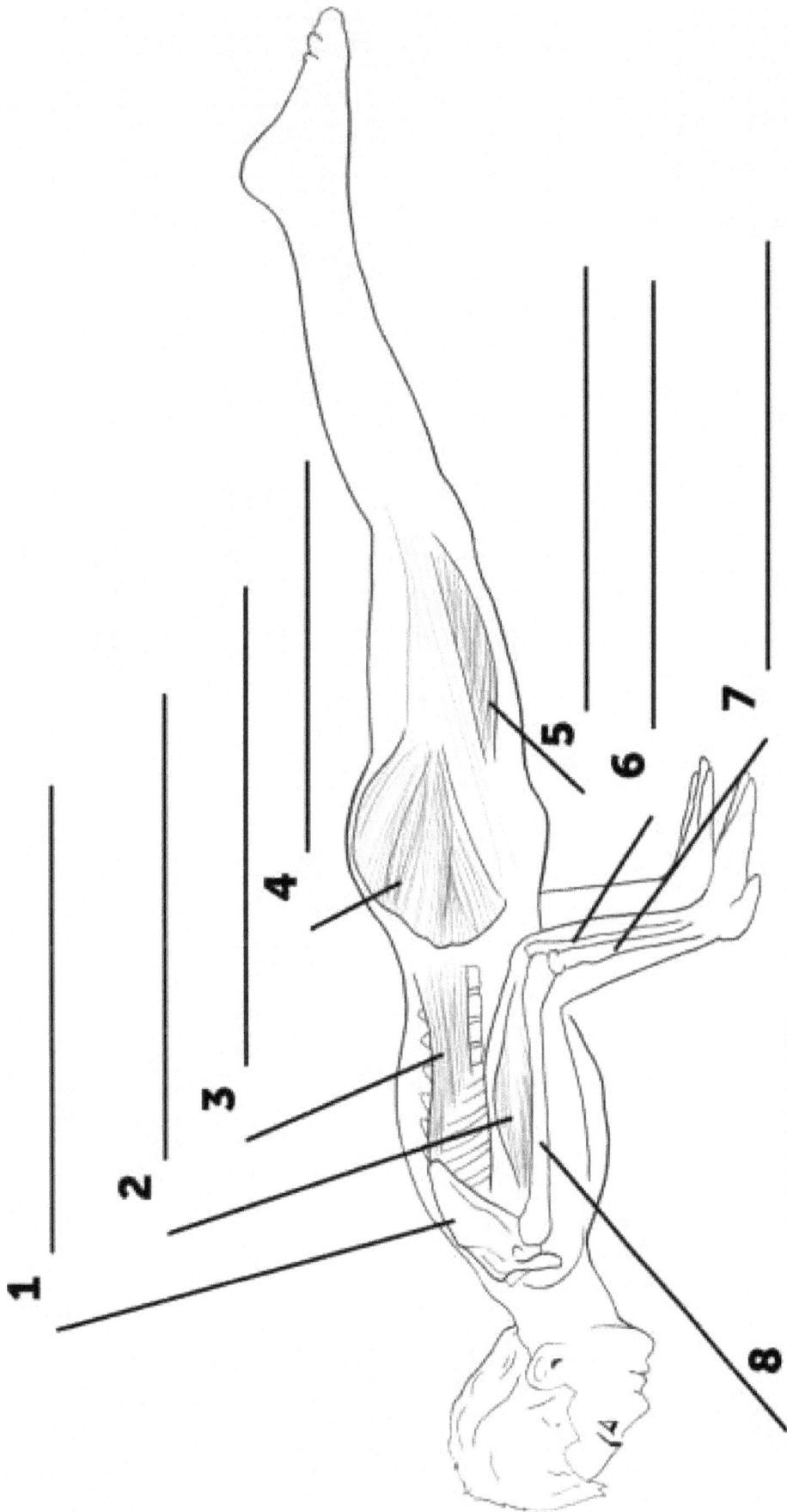

113. PEACOCK POSE

1. SCAPULA
2. TRICEPS BRACHII
3. ERECTOR SPINAE
4. GLUTEUS MAXIMUS
5. QUADRICEPS
6. ULNA
7. RADIUS
8. HUMERUS

114. ONE-LEGGED KING PIGEON POSE II

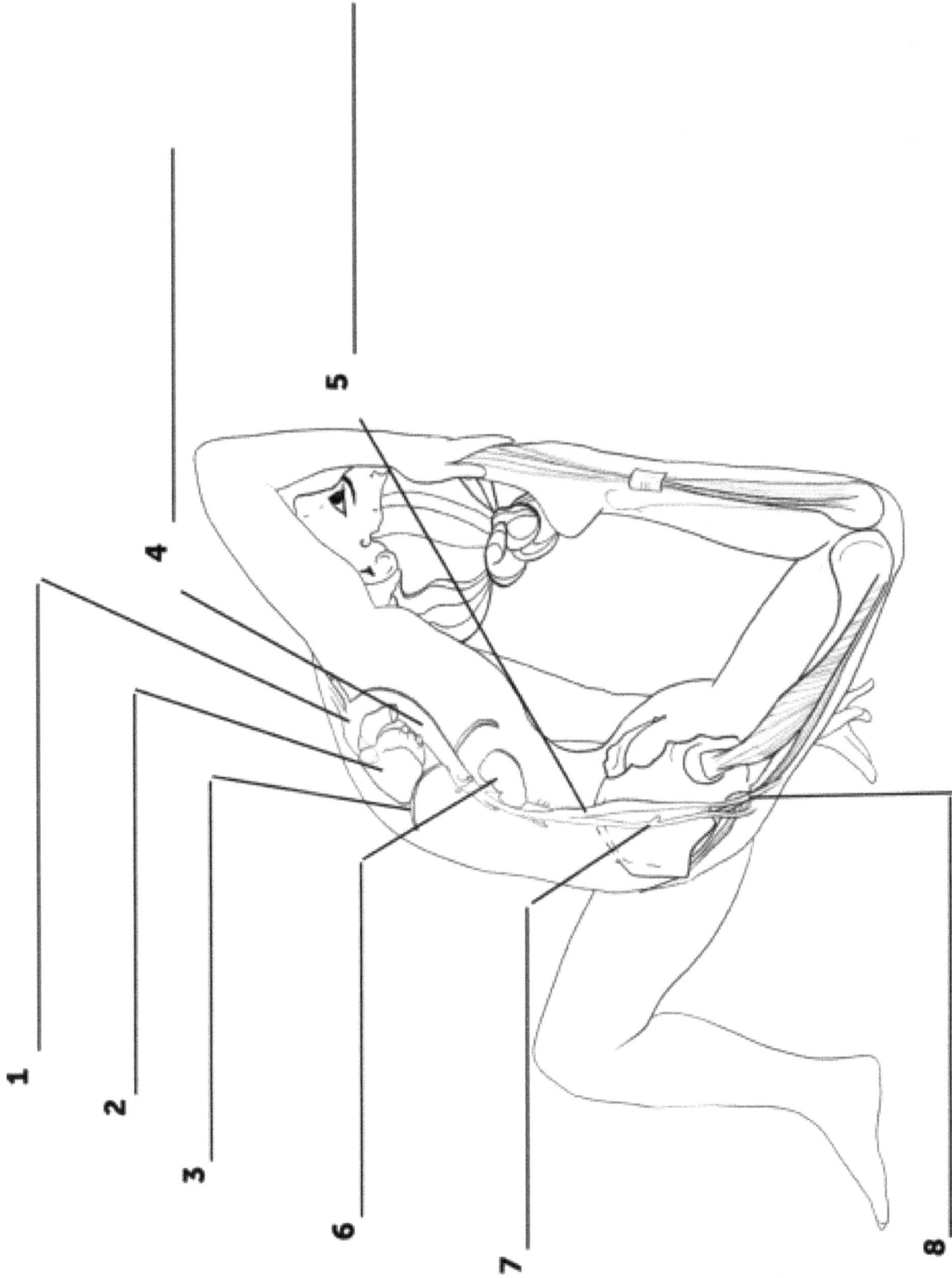

114. ONE-LEGGED KING PIGEON POSE II

1. ASCENDING THORACIC AORTA
2. HEART
3. DIAPHRAGM
4. DESCENDING THORACIC AORTA
5. ABDOMINAL AORTA
6. KIDNEY
7. COMMON ILIAC ARTERY
8. FEMORAL ARTERY

115. LITTLE THUNDERBOLT POSE

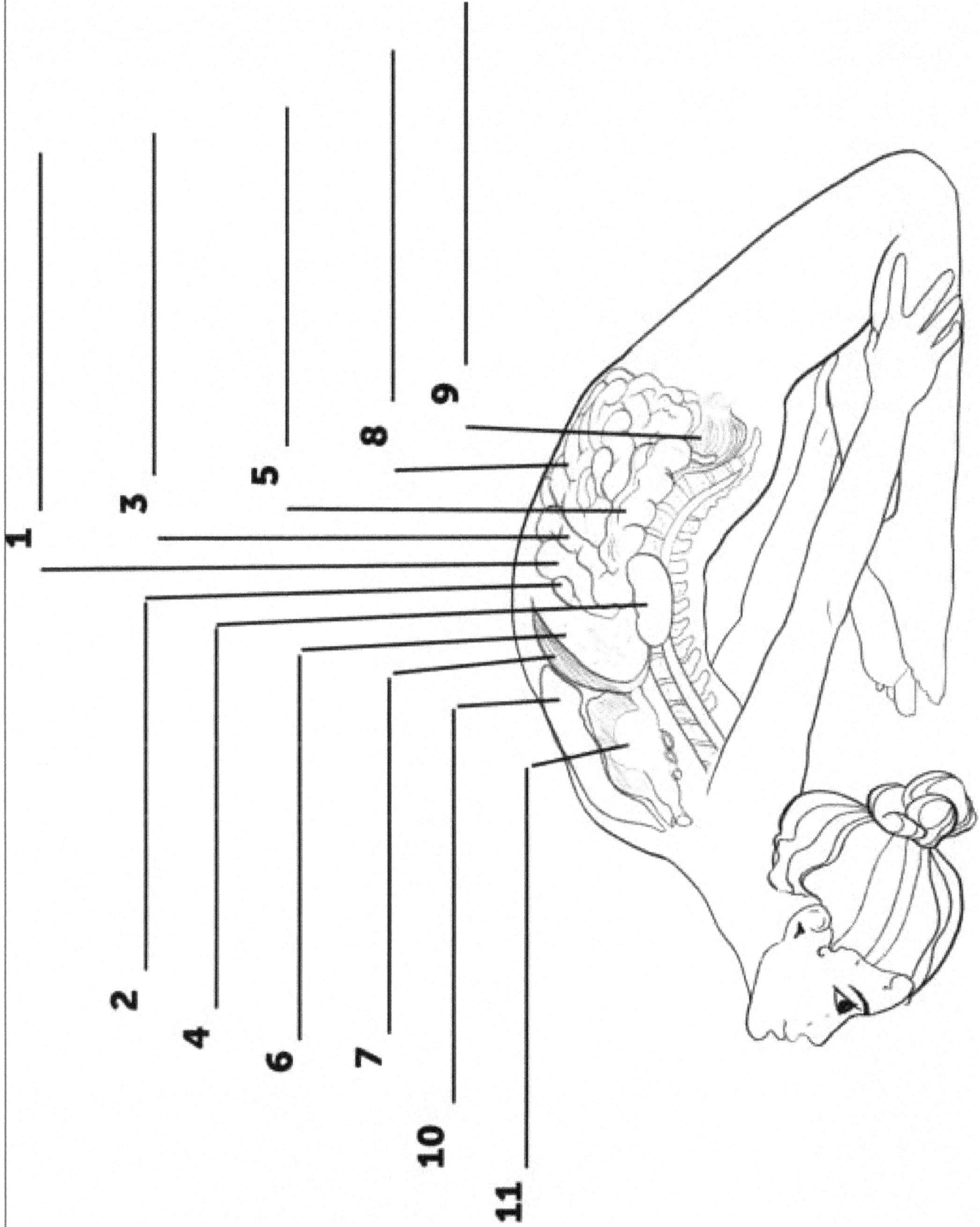

115. LITTLE THUNDERBOLT POSE

1. STOMACH
2. GALLBLADDER
3. TRANSVERSE COLON
4. KIDNEY
5. ASCENDING COLON
6. LIVER
7. DIAPHRAGM
8. COILS OF SMALL INTESTINE
9. RECTUM
10. LUNGS
11. HEART

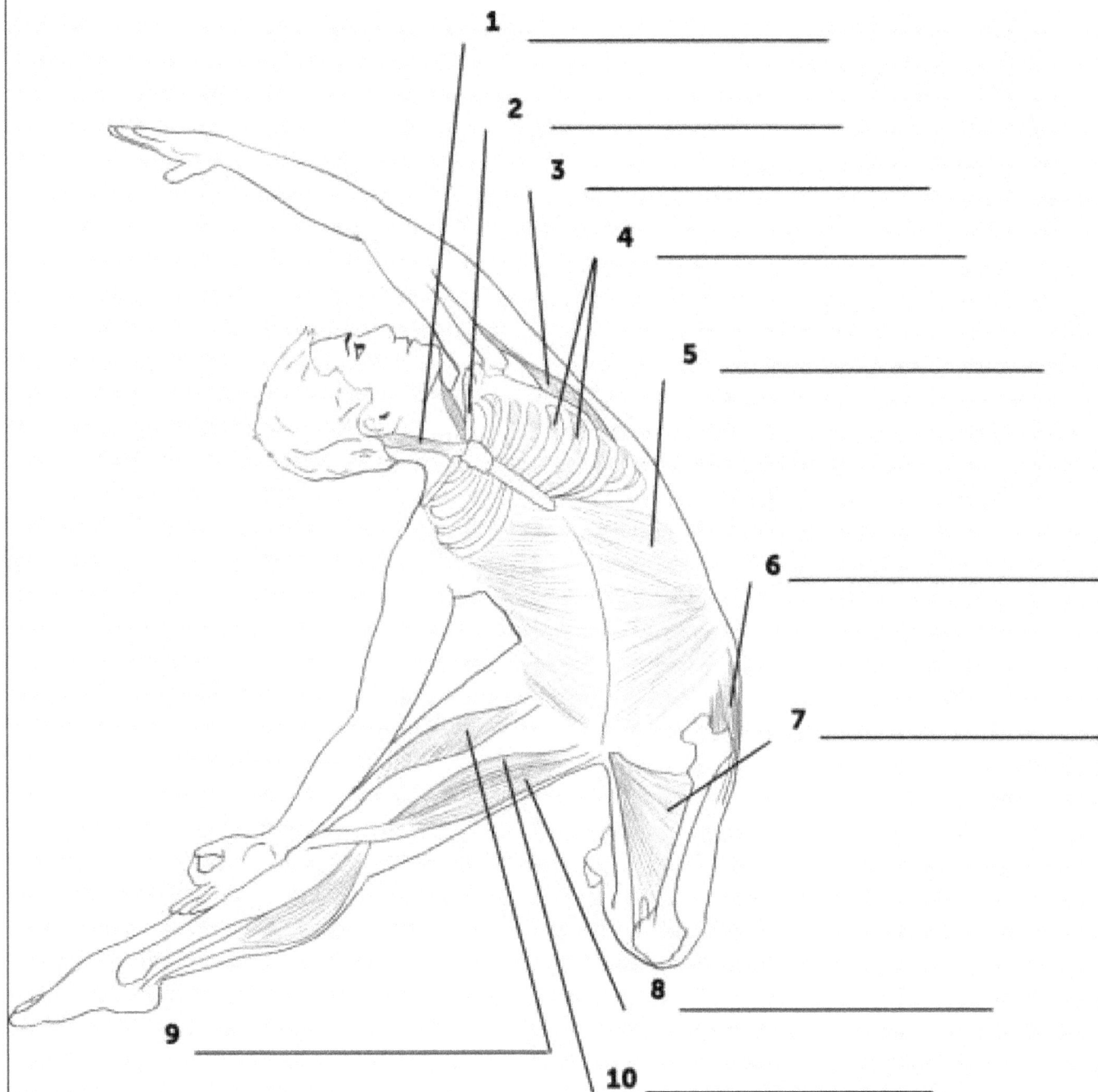

116. GATE POSE

1 _____

2 _____

3 _____

4 _____

5 _____

6 _____

7 _____

8 _____

9 _____

10 _____

116. GATE POSE

1. SPLENIUS CAPITIS
2. COLLARBONE
3. LATISSIMUS DORSI
4. INTERCOSTALS
5. EXTERNAL OBLIQUE
6. TENSOR FASCIAE LATAE
7. ADDUCTOR LONGUS
8. GRACILIS
9. RECTUS FEMORIS
10. ADDUCTOR MAGNUS

117. SAGE KOUNDIYA I POSE

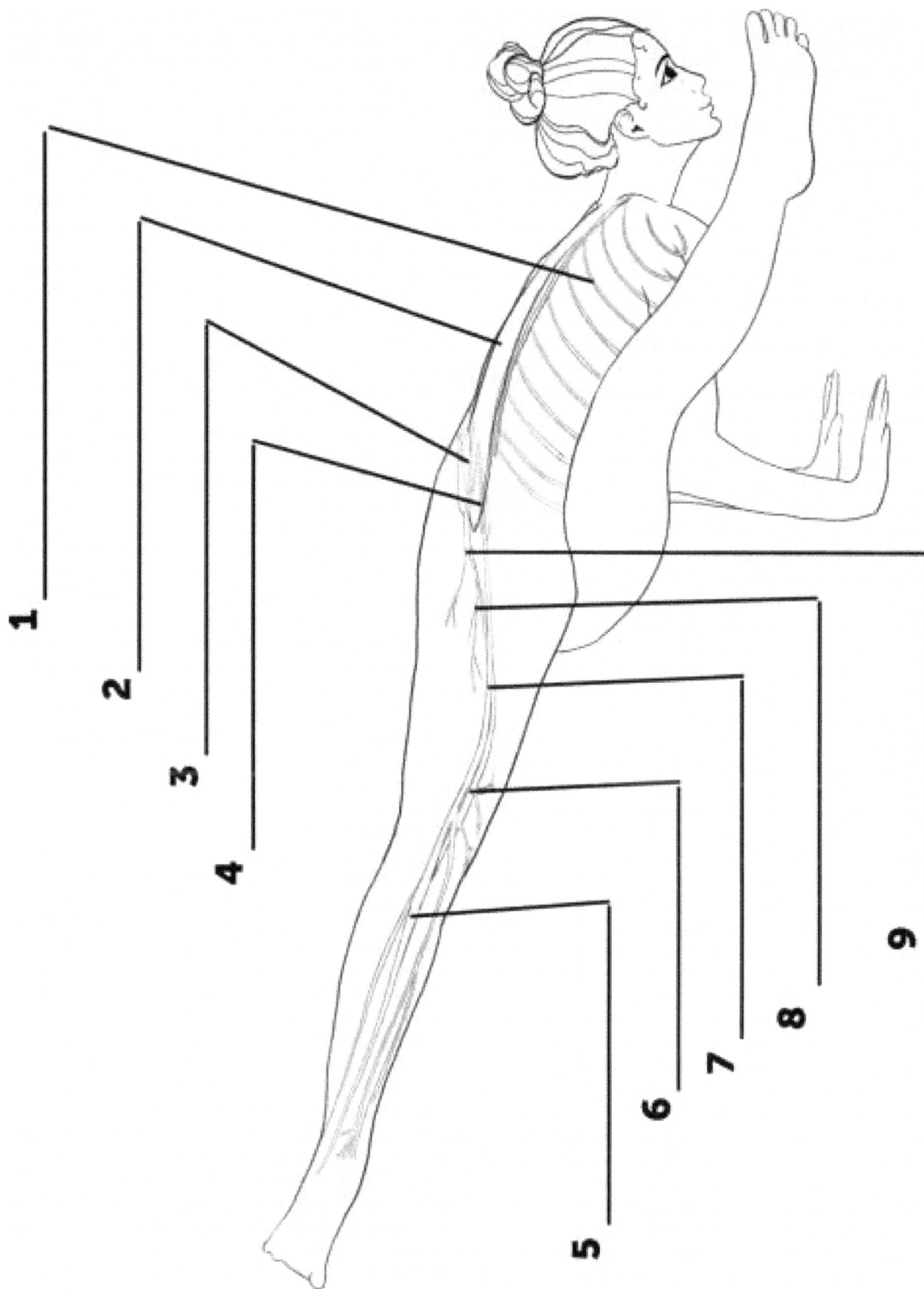

1

2

3

4

5

6

7

8

9

117. SAGE KOUNDIYA I POSE

1. INTERCOSTALS

2. SPINAL CORD

3. LUMBAR PLEXUS

4. SACRAL PLEXUS

5. TIBIAL

6. SAPHENOUS

7. SCIATIC

8. MUSCULAR BRANCHES OF FEMORAL

9. FEMORAL

118. SAGE KOUNDIYA II POSE

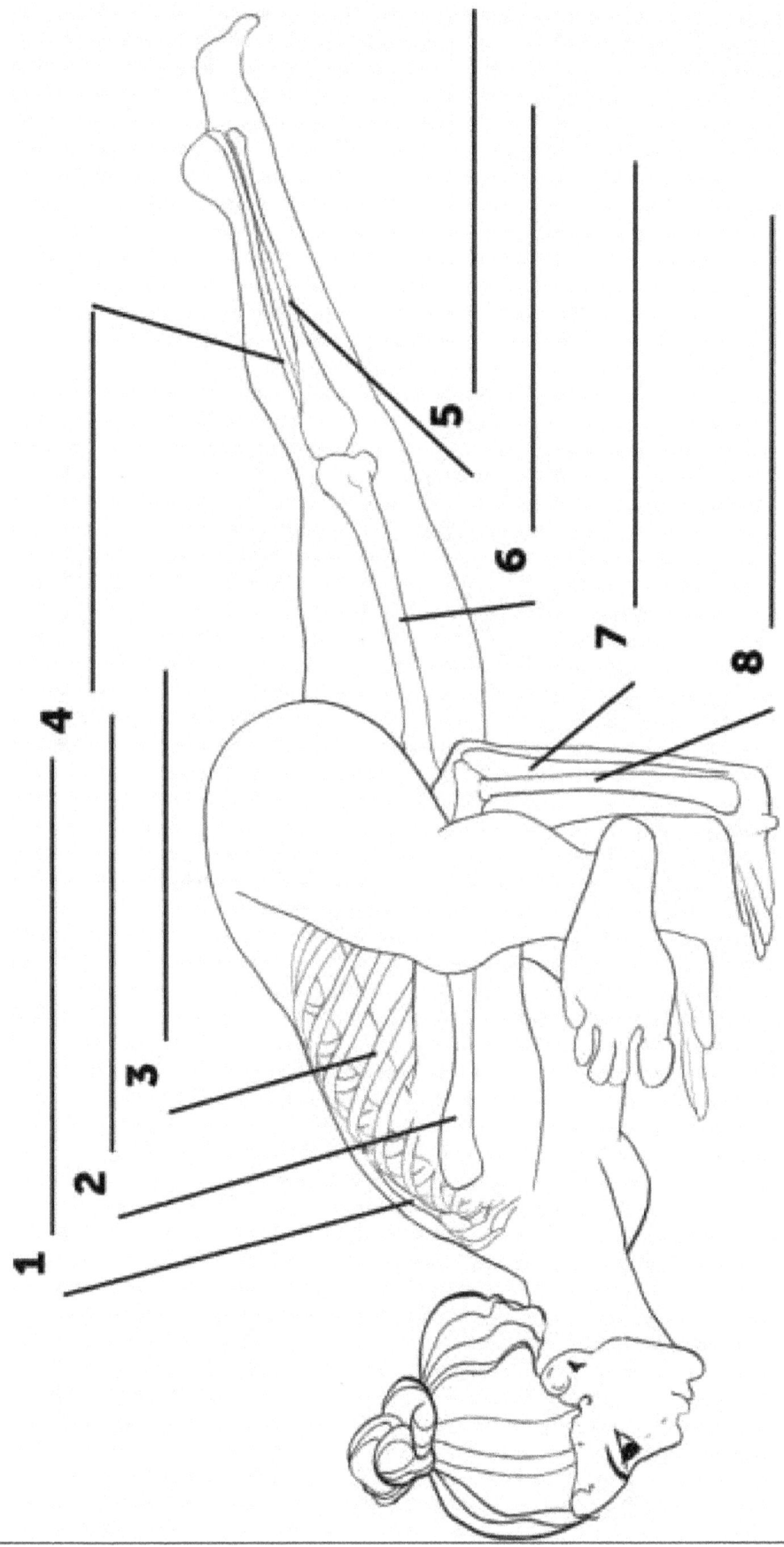

1

2

3

4

5

6

7

8

118. SAGE KOUNDIYA II POSE

1. SCAPULA

2. HUMERUS

3. RIBS

4. FIBULA

5. TIBIA

6. FEMUR

7. ULNA

8. RADIUS

119. HEAD TO FOOT POSE

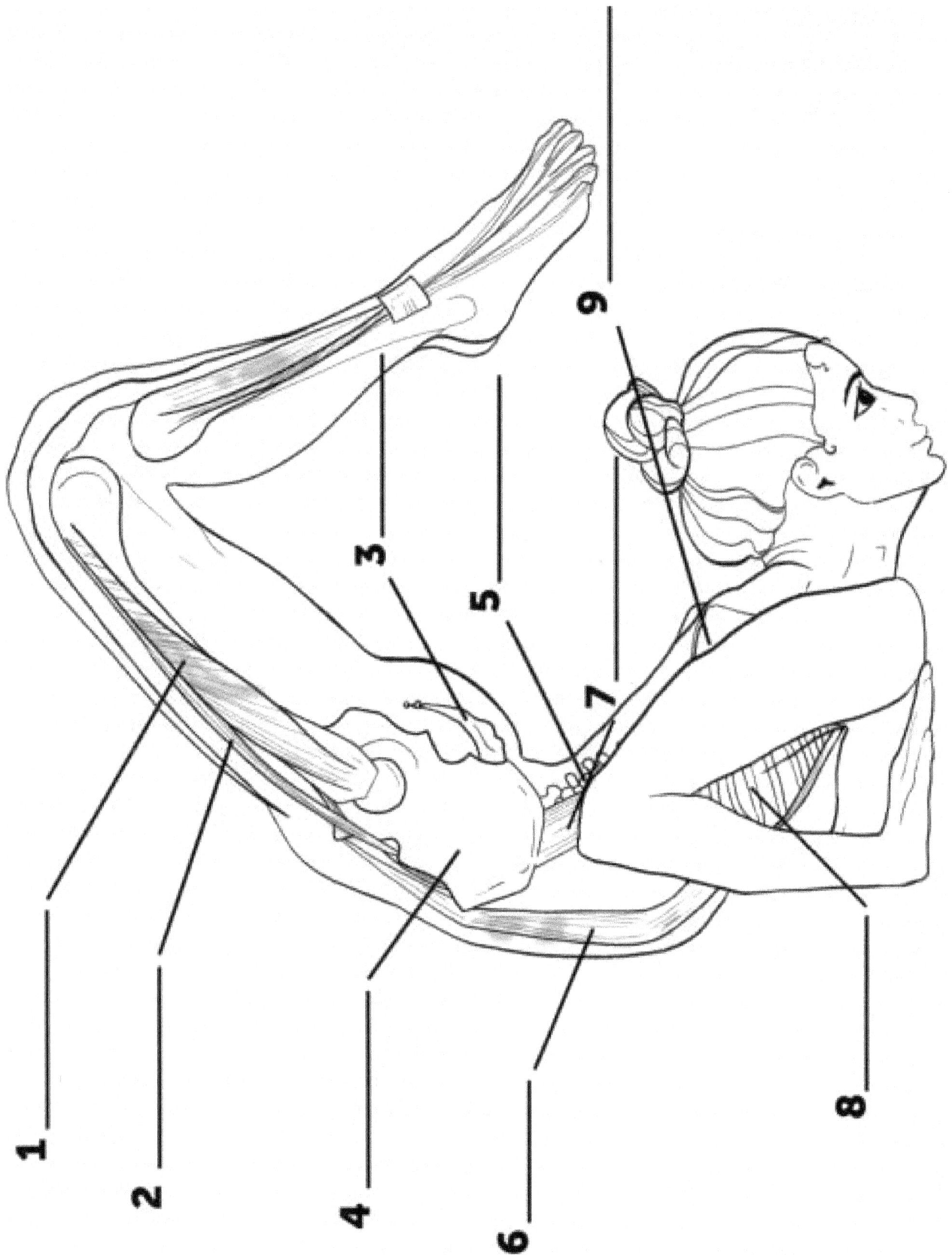

1

2

3

4

5

6

7

8

9

119. HEAD TO FOOT POSE

1. VASTUS LATERALIS
2. RECTUS FEMORIS
3. SACRUM
4. PELVIS
5. SPINE
6. RECTUS ABDOMINIS
7. ERECTOR SPINAE
8. RIBS
9. SCAPULA

120. MASTER BABY GRASSHOPPER POSE

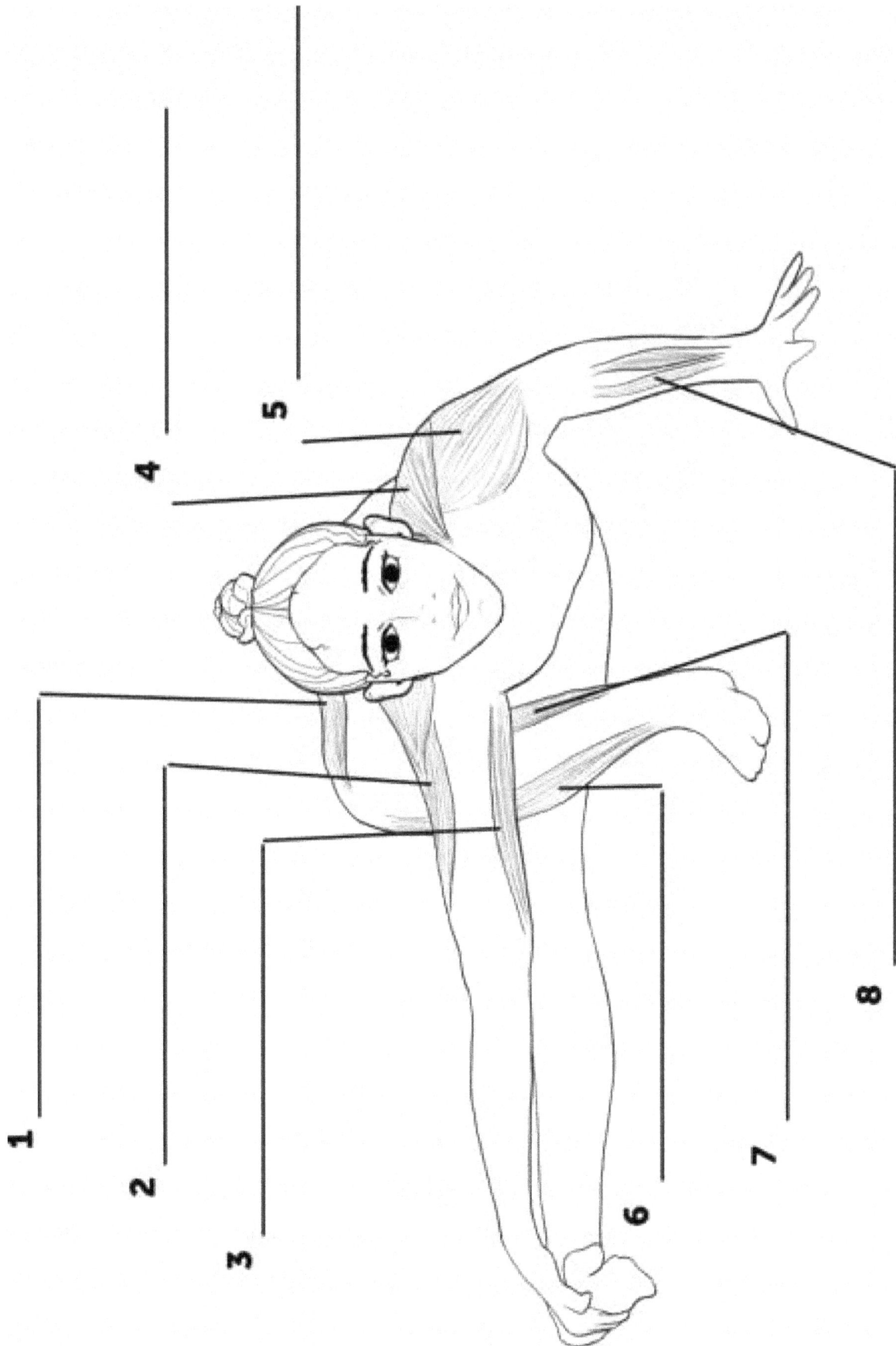

120. MASTER BABY GRASSHOPPER POSE

1. QUADRICEPS
2. TRICEPS BRACHII
3. BICEPS BRACHII
4. TRAPEZIUS
5. DELTOID
6. TIBIALIS ANTERIOR
7. GASTROCNEMIUS
8. PRONATORS

121. UPWARD-FACING TWO-FOOTED STAFF POSE

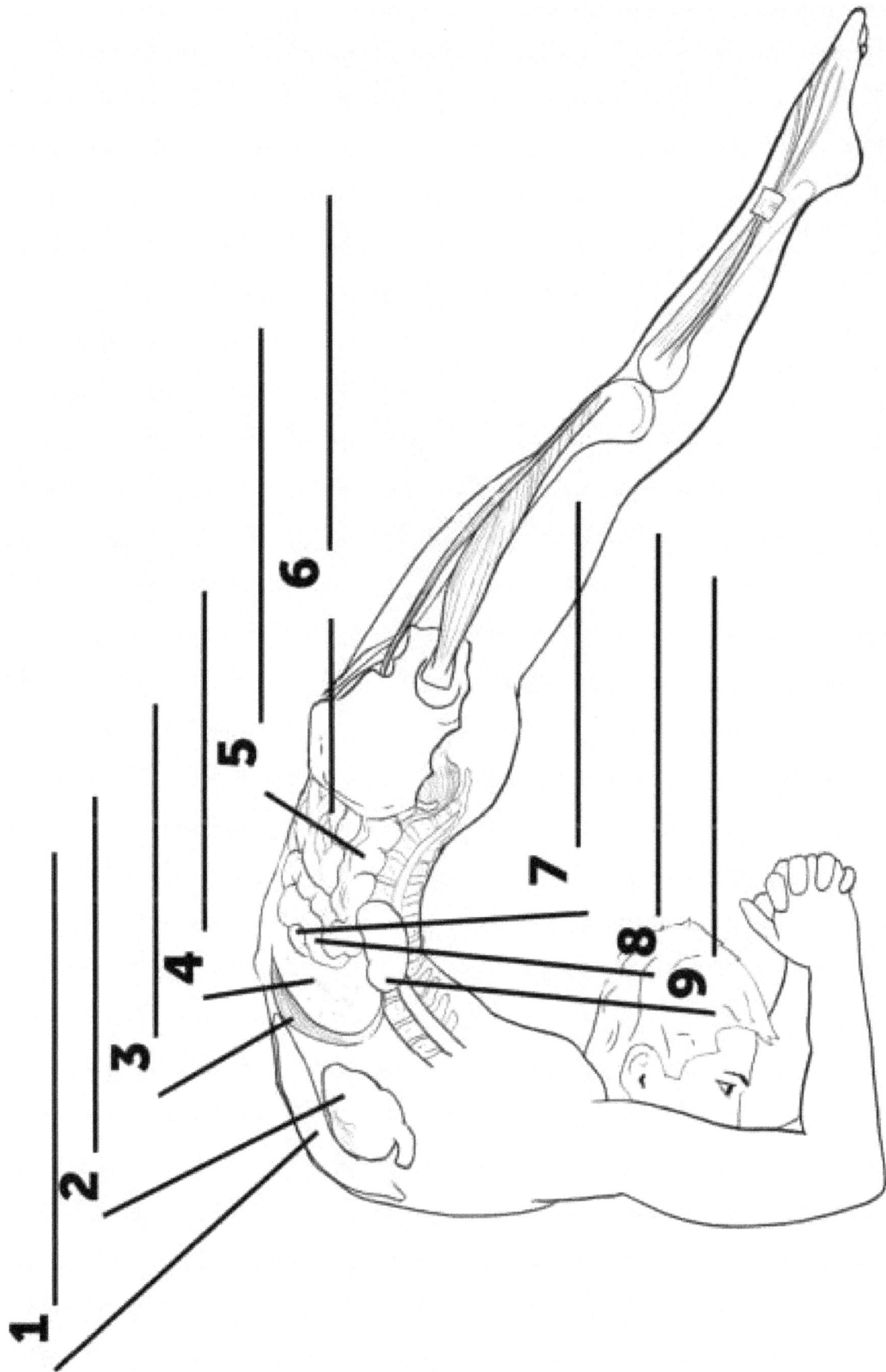

121. UPWARD-FACING TWO-FOOTED STAFF POSE

1. LUNGS
2. HEART
3. DIAPHRAGM
4. LIVER
5. ASCENDING COLON
6. COILS OF SMALL INTESTINE
7. GALLBLADDER
8. STOMACH
9. KIDNEY

122. BHARADVAJA'S TWIST

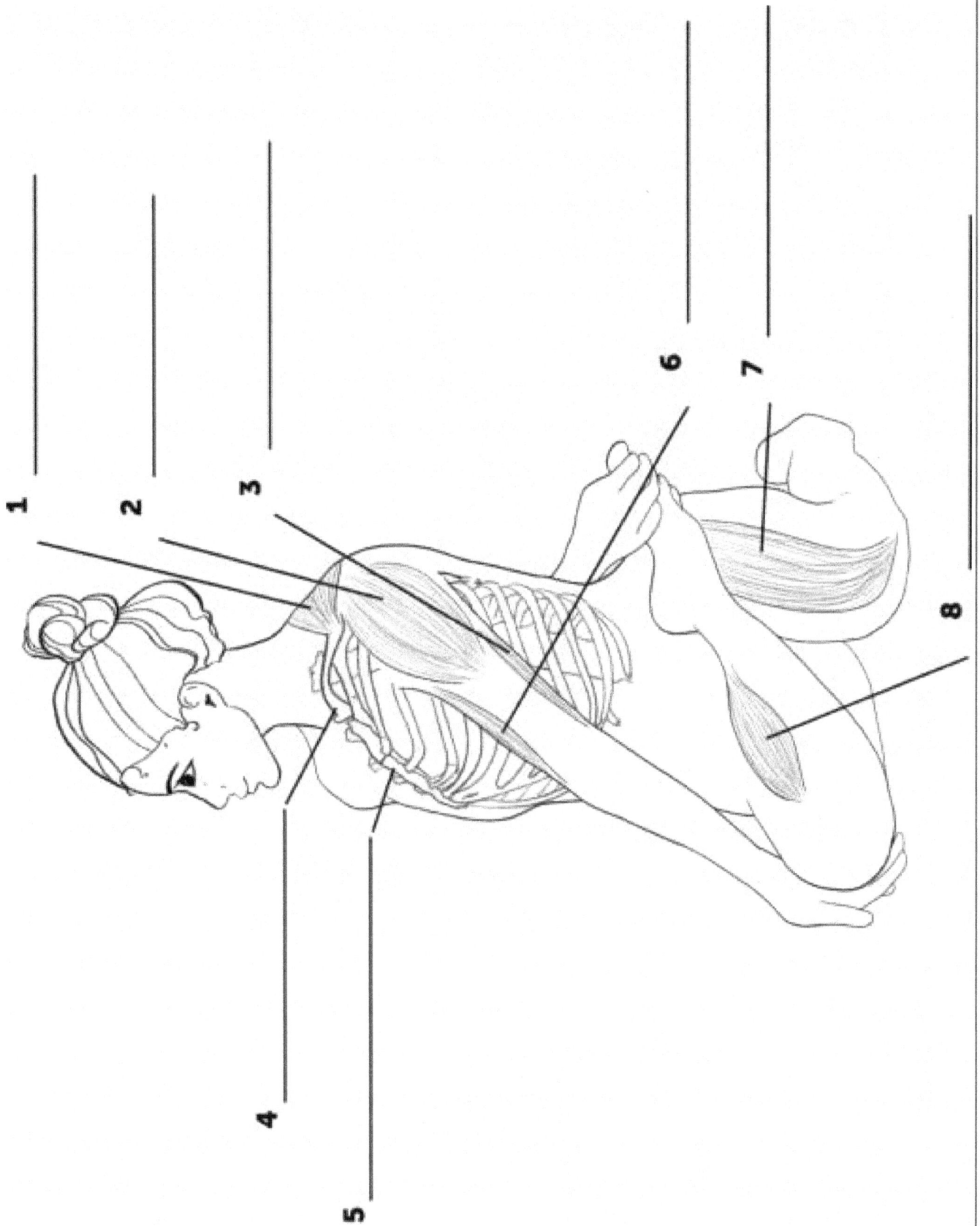

1

2

3

4

5

6

7

8

122. BHARADVAJA'S TWIST

1. TRAPEZIUS
2. DELTOID
3. TRICEPS BRACHII
4. COLLARBONE
5. STERNUM
6. BICEPS BRACHII
7. QUADRICEPS
8. GASTROCNEMIUS

123. EIGHT ANGLE POSE

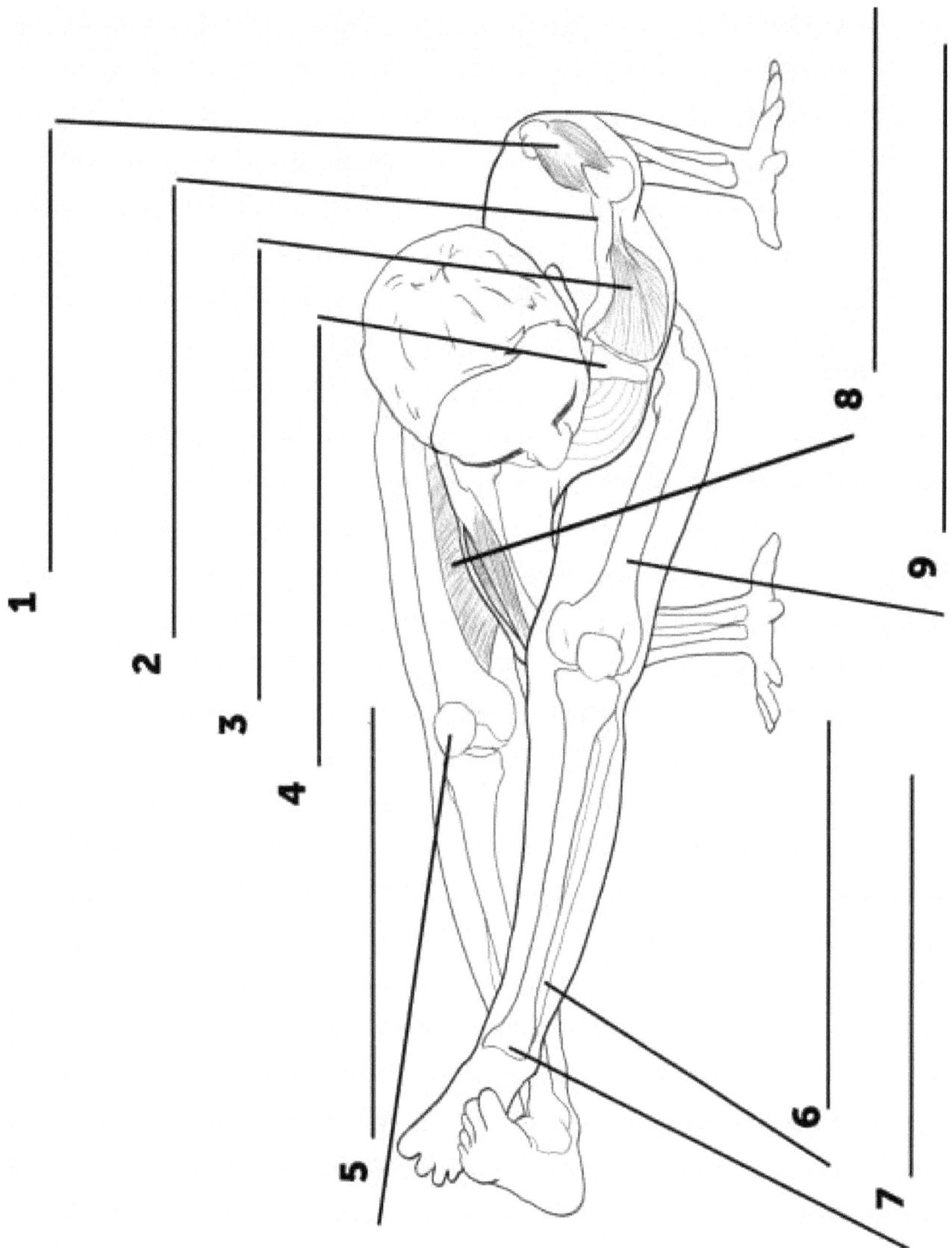

1

2

3

4

5

6

7

8

9

123. EIGHT ANGLE POSE

1. TRICEPS BRACHII

2. COLLARBONE

3. PECTORALIS MAJOR

4. STERNUM

5. PATELLA

6. FIBULA

7. TIBIA

8. ADDUCTORS

9. FEMUR

124. SAGE HALF BOUND LOTUS POSE

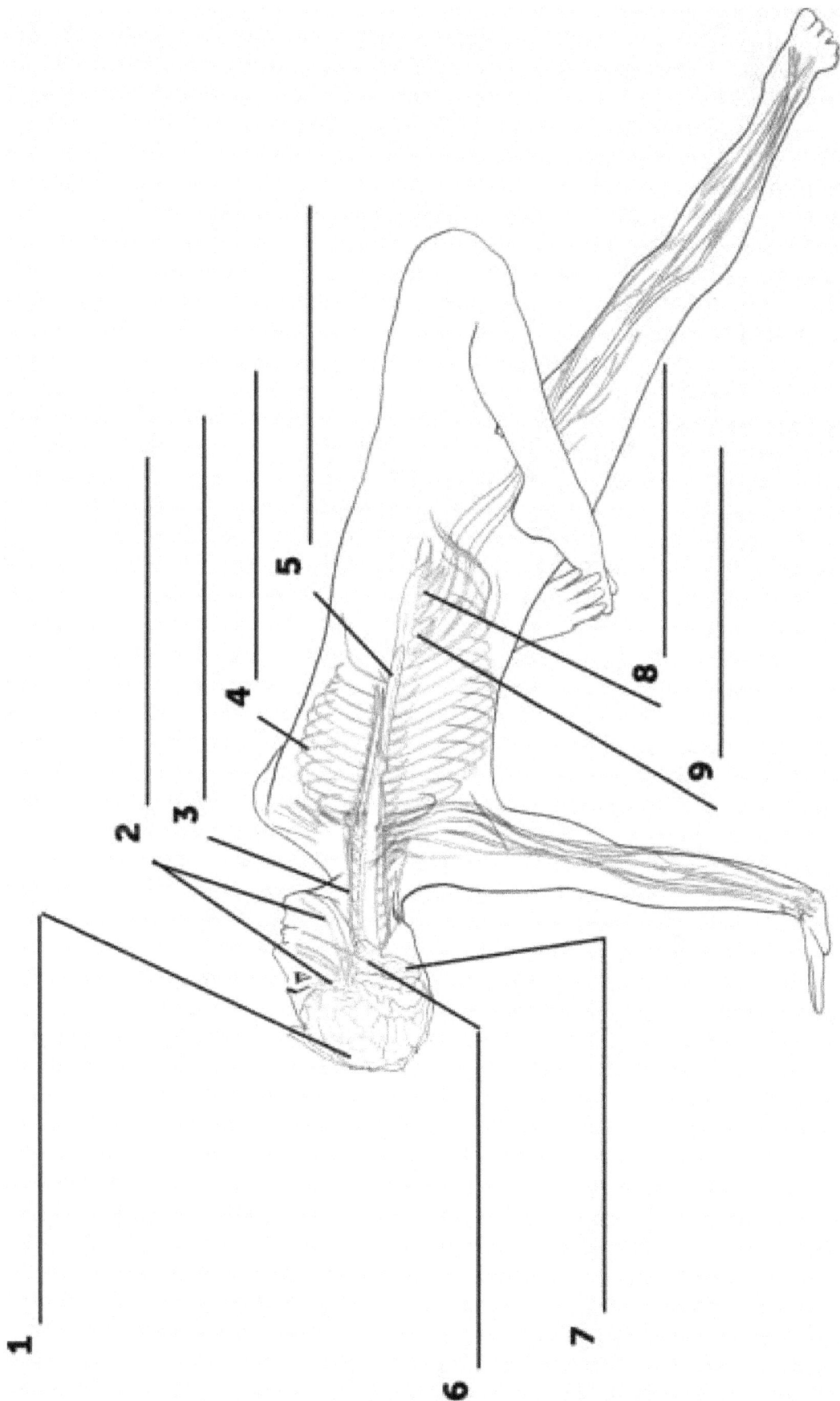

1

2

3

4

5

6

7

8

9

124. SAGE HALF BOUND LOTUS POSE

1. CEREBRUM
2. CRANIAL NERVES
3. VAGUS
4. INTERCOSTALS
5. SPINAL CORD
6. BRAINSTEM
7. CEREBELLUM
8. SACRAL PLEXUS
9. LUMBAR PLEXUS

125. SHOULDER PRESSING POSE

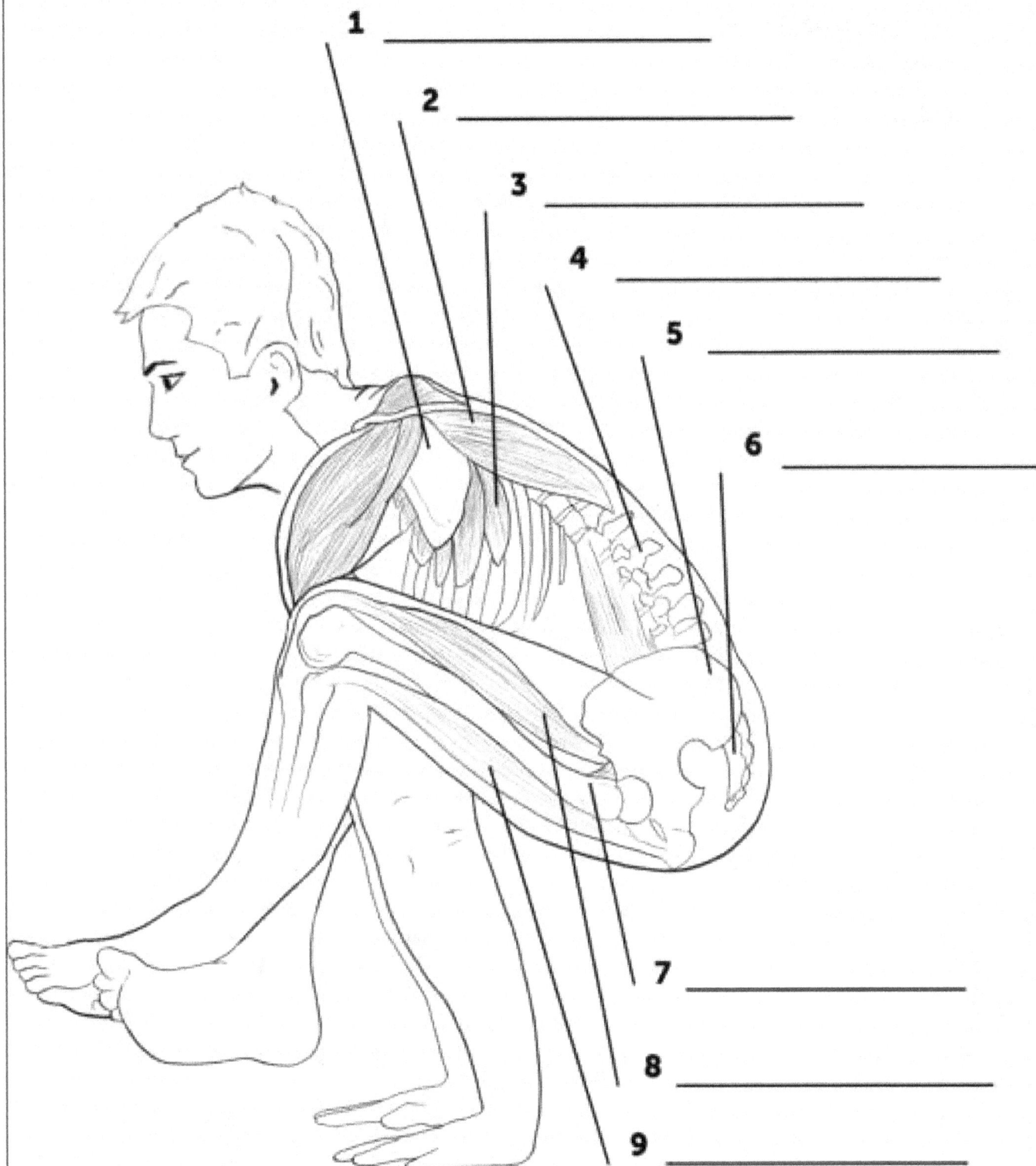

1 _____

2 _____

3 _____

4 _____

5 _____

6 _____

7 _____

8 _____

9 _____

125. SHOULDER PRESSING POSE

1. SCAPULA
2. RHOMBOIDS
3. SERRATUS ANTERIOR
4. SPINE
5. PELVIS
6. SACRUM
7. FEMUR
8. QUADRICEPS
9. HAMSTRINGS

126. SUPER SOLDIER

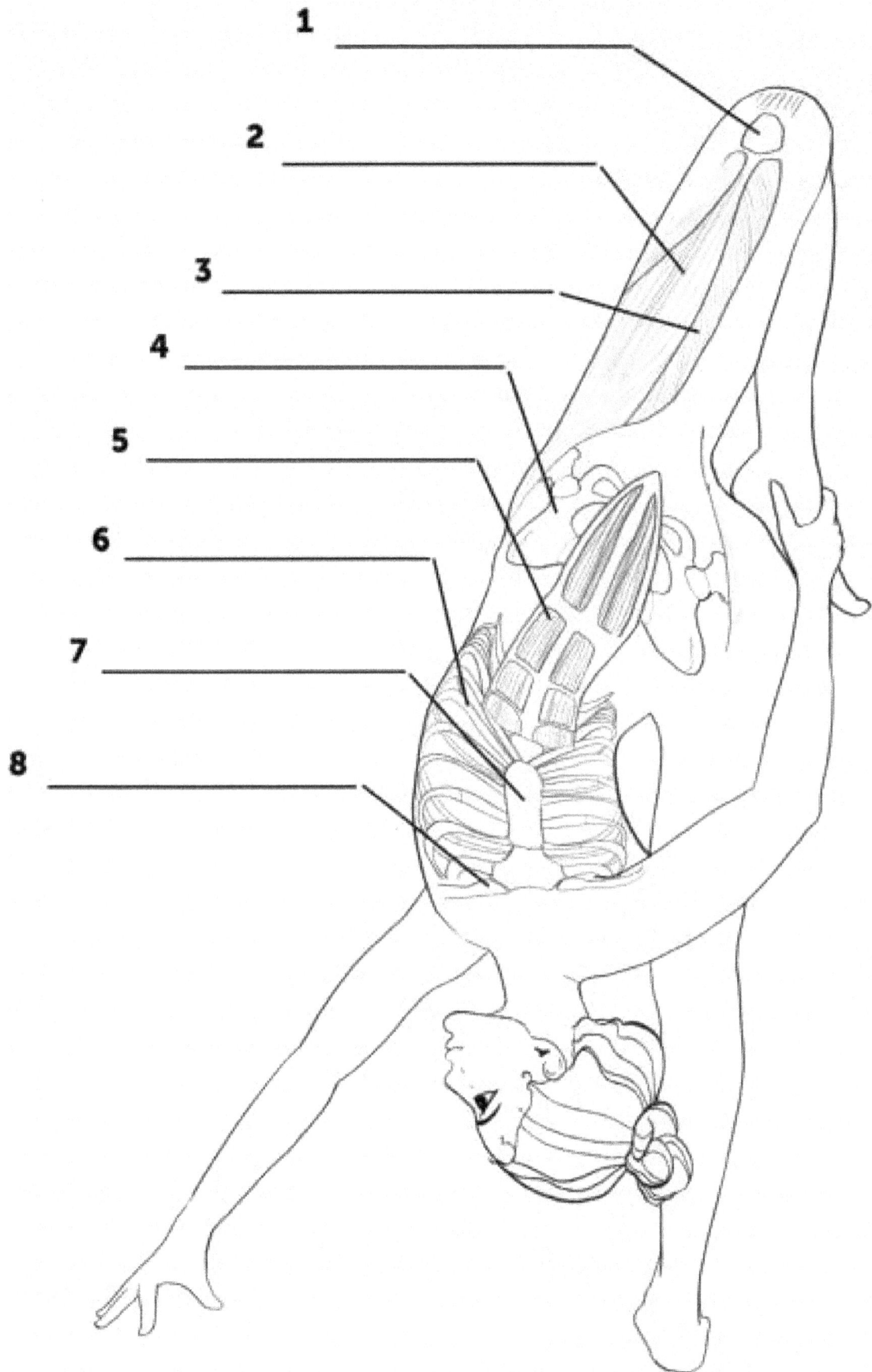

1 _____

2 _____

3 _____

4 _____

5 _____

6 _____

7 _____

8 _____

126. SUPER SOLDIER

1.   PATELLA

2.   RECTUS FEMORIS

3.   VASTUS MEDIALIS

4.   PELVIS

5.   RECTUS ABDOMINIS

6.   RIBS

7.   STERNUM

8.   COLLARBONE

127. MONKEY POSE

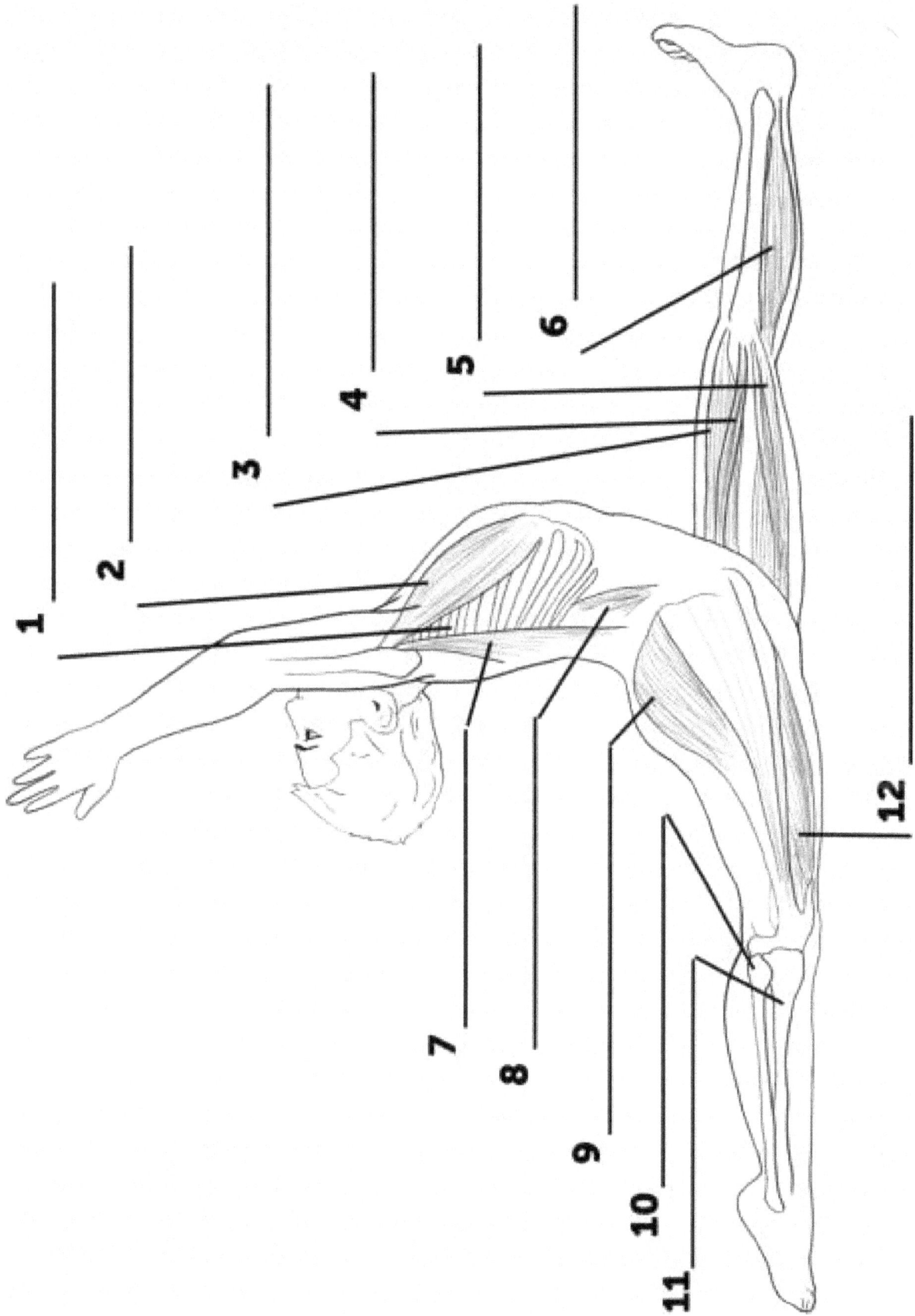

127. MONKEY POSE

1. RIBS
2. PECTORALIS MAJOR
3. RECTUS FEMORIS
4. SARTORIUS
5. HAMSTRINGS
6. GASTROCNEMIUS
7. LATISSIMUS DORSI
8. ERECTOR SPINAE
9. GLUTEUS MAXIMUS
10. FIBULA
11. TIBIA
12. QUADRICEPS

128. SEATED WIDE ANGLE POSE

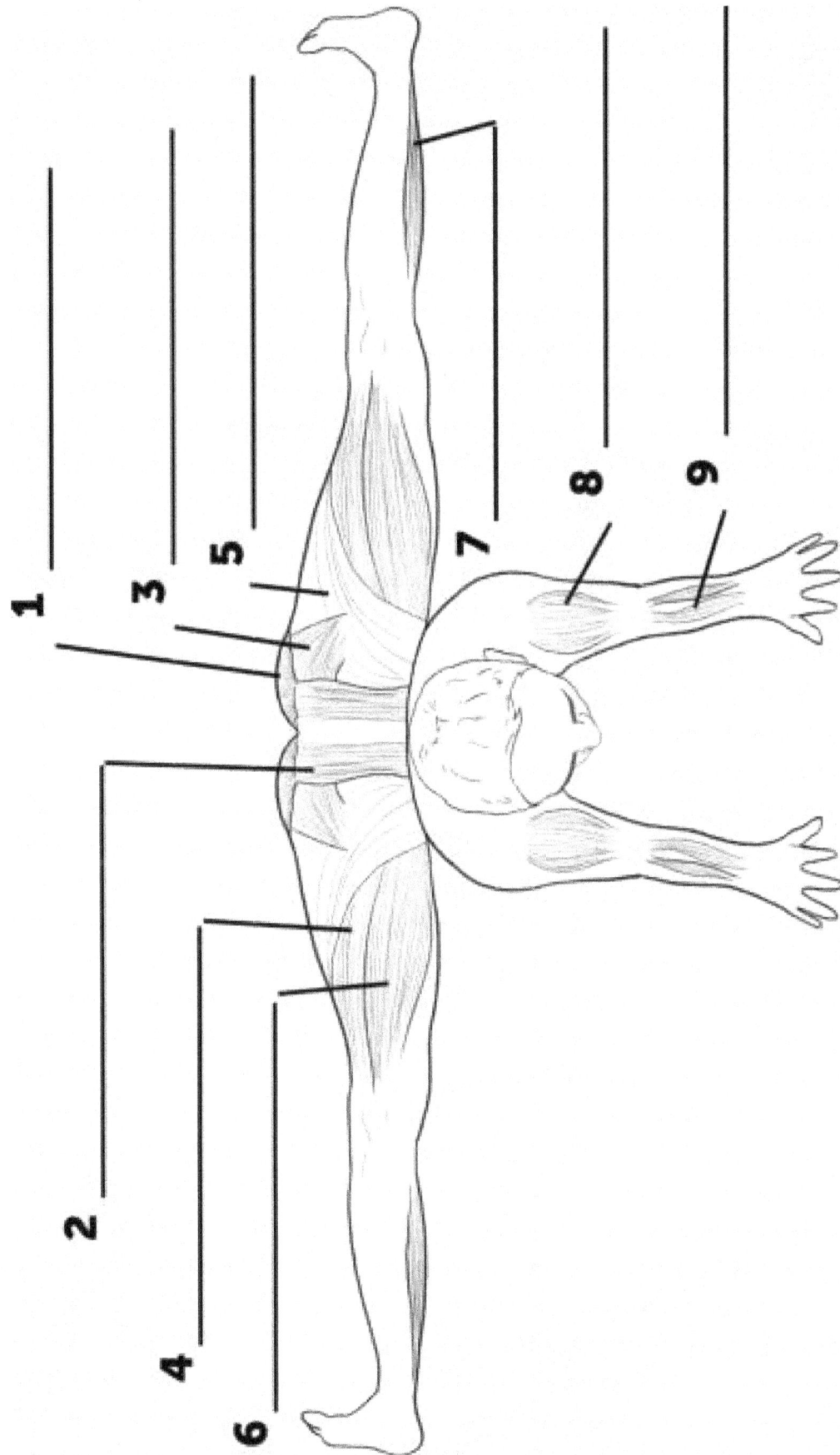

1

2

3

4

5

6

7

8

9

128. SEATED WIDE ANGLE POSE

1. GLUTEUS MAXIMUS

2. ERECTOR SPINAE

3. GLUTEUS MEDIUS

4. VASTUS LATERALIS

5. ILIOTIBIAL BAND

6. RECTUS FEMORIS

7. GASTROCNEMIUS

8. DELTOID

9. PRONATORS

129. EXTENDED BALANCING LIZARD

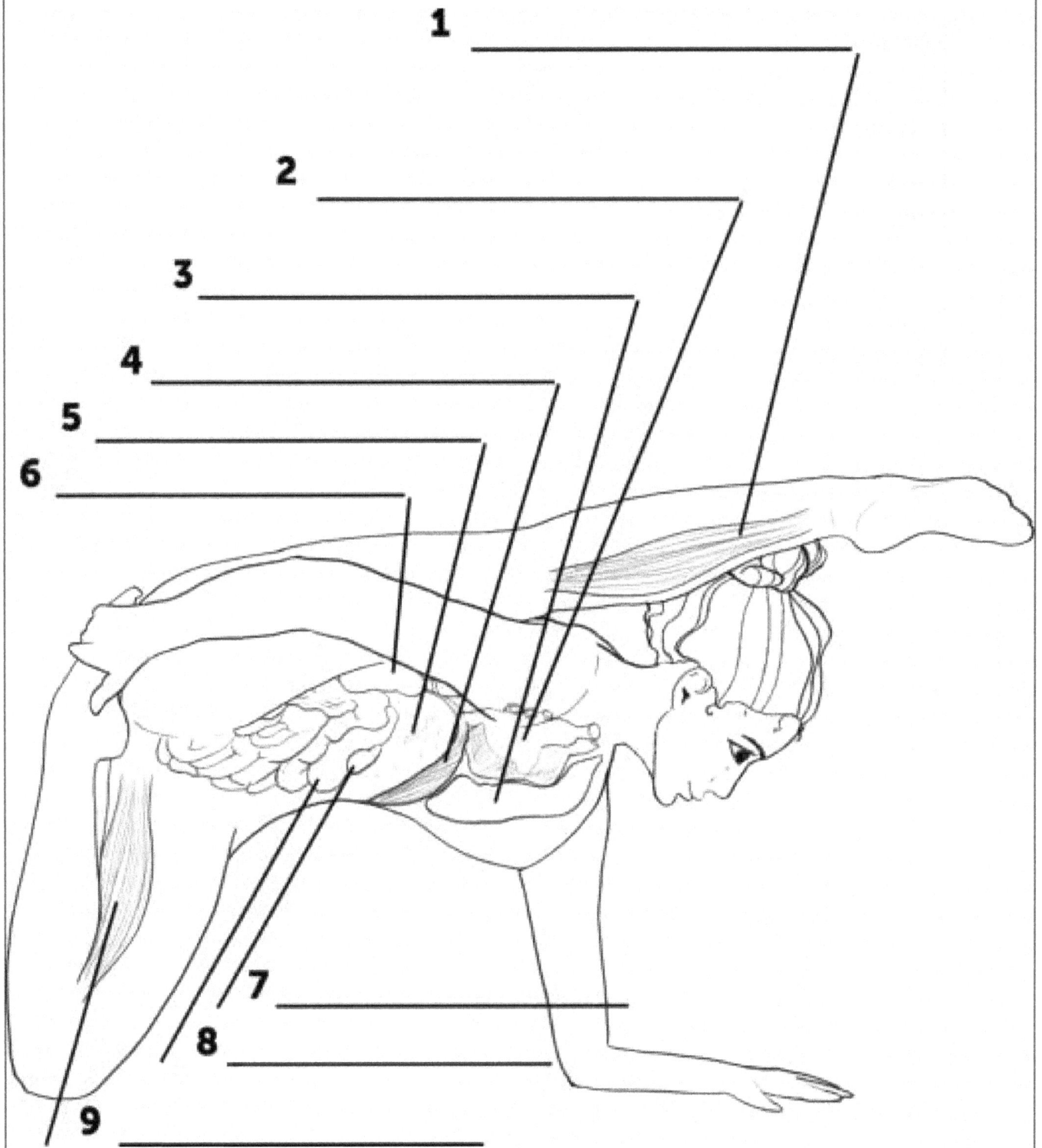

1 _____

2 _____

3 _____

4 _____

5 _____

6 _____

7 _____

8 _____

9 _____

129. EXTENDED BALANCING LIZARD

1. GASTROCNEMIUS
2. HEART
3. LUNGS
4. DIAPHRAGM
5. LIVER
6. KIDNEY
7. GALLBLADDER
8. STOMACH
9. HAMSTRINGS

130. KURMASANA

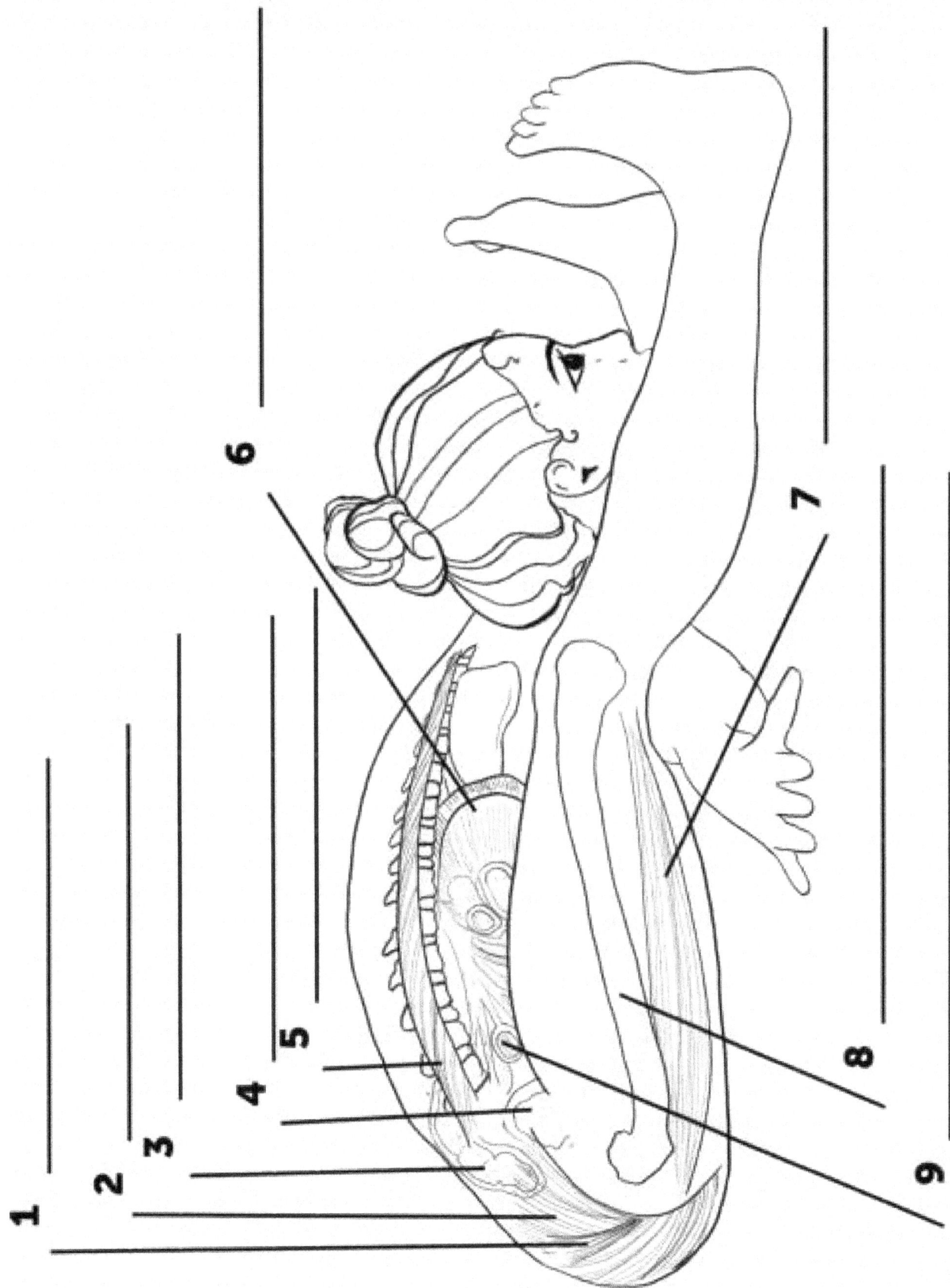

130. KURMASANA

1. PIRIFORMIS
2. GLUTEUS MAXIMUS
3. RECTUM
4. URINARY BLADDER
5. SPINAL MUSCLES
6. DIAPHRAGM
7. HAMSTRINGS
8. FEMUR
9. COILS OF SMALL INTESTINE

131. VIPARITA SALABHASANA

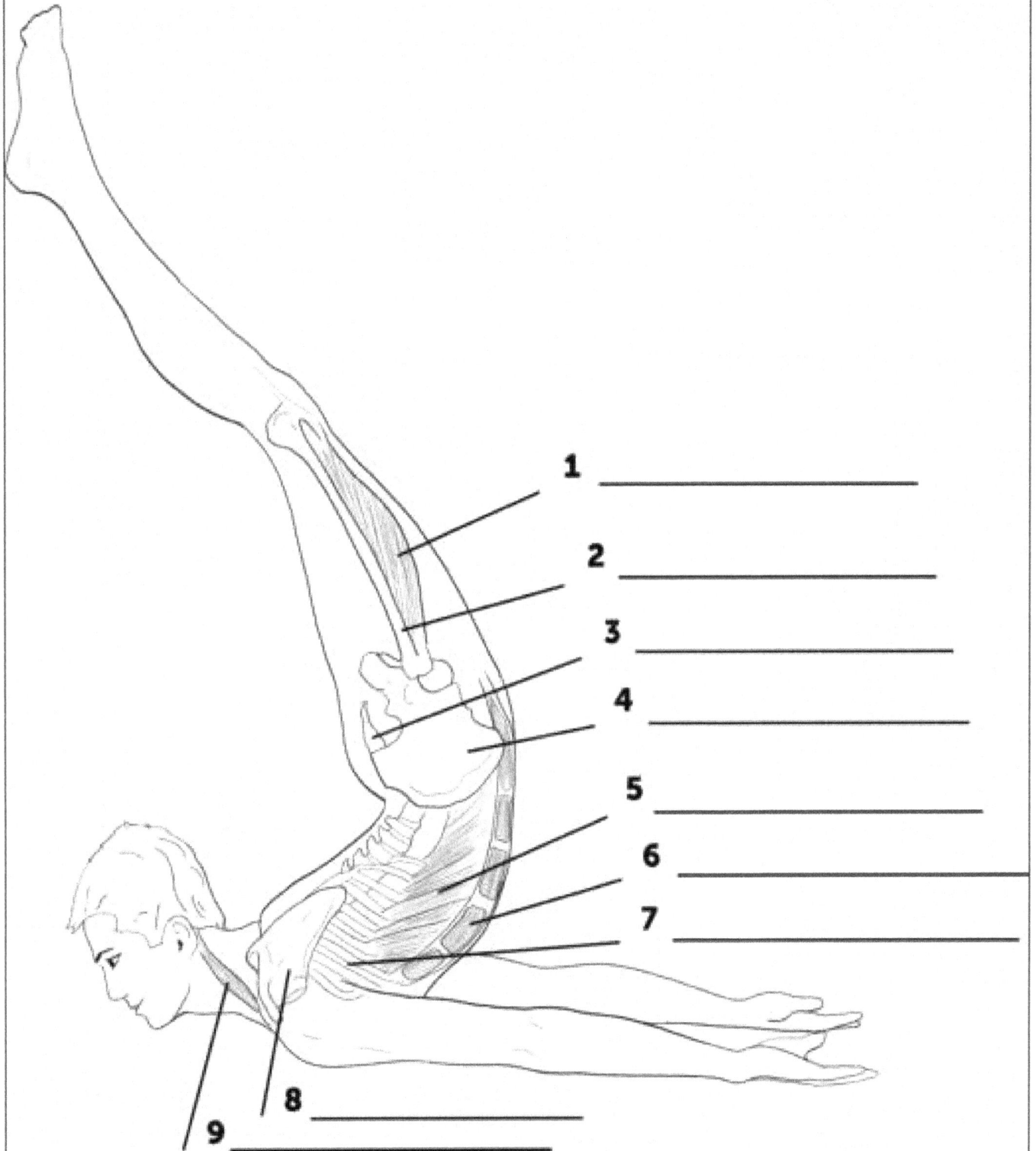

1 _____

2 _____

3 _____

4 _____

5 _____

6 _____

7 _____

8 _____

9 _____

131. VIPARITA SALABHASANA

1. QUADRICEPS
2. FEMUR
3. SACRUM
4. PELVIS
5. EXTERNAL OBLIQUE
6. RECTUS ABDOMINIS
7. RIBS
8. SCAPULA
9. STERNOCLEIDOMASTOID

132. SLEEPING YOGI POSE

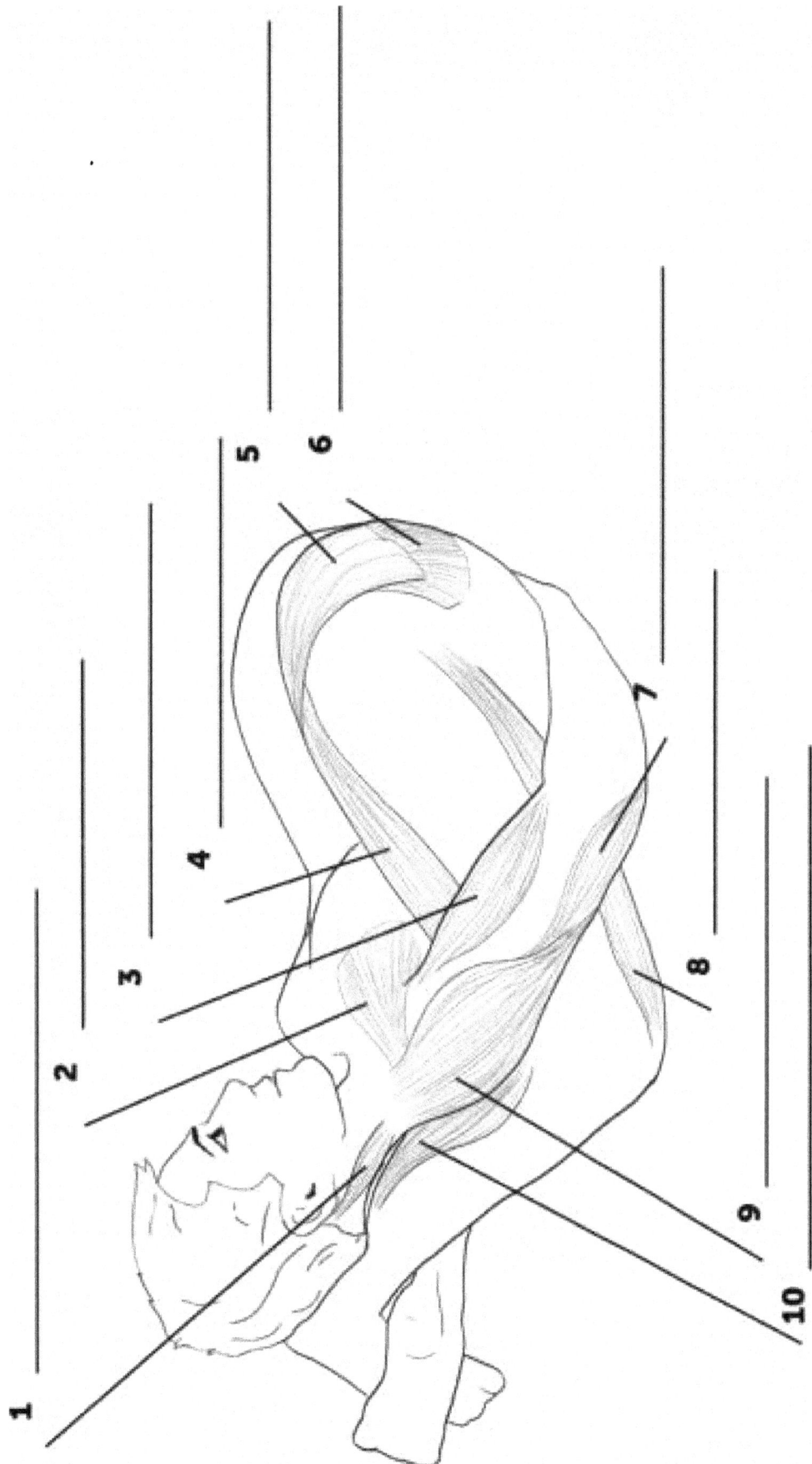

132. SLEEPING YOGI POSE

1. STERNOCLEIDOMASTOID
2. PECTORALIS MAJOR
3. BICEPS BRACHII
4. HAMSTRINGS
5. GLUTEUS MAXIMUS
6. GLUTEUS MEDIUS
7. TRICEPS BRACHII
8. QUADRICEPS
9. DELTOID
10. GASTROCNEMIUS

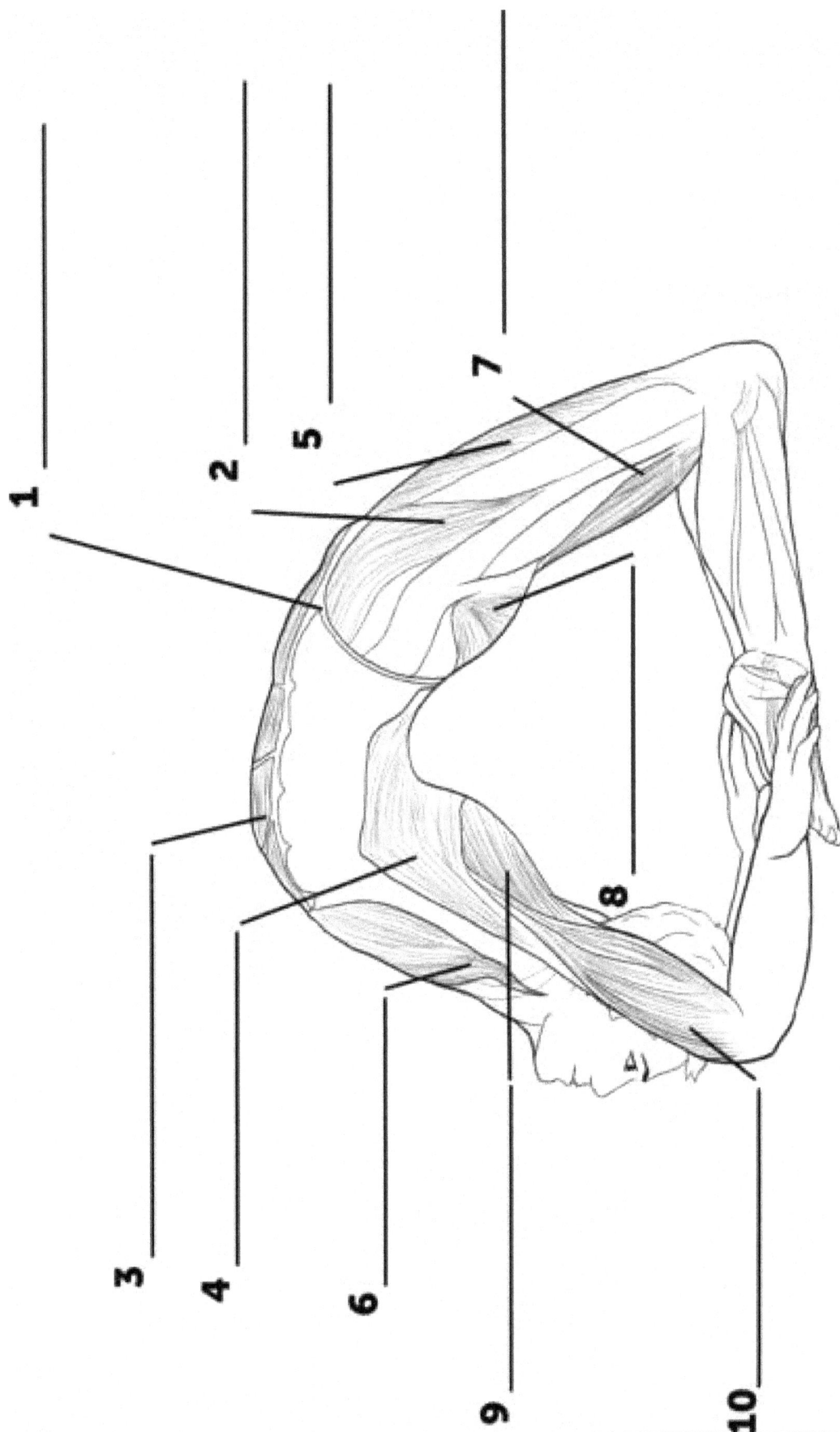

133. DOVE POSE

1

2

3

4

5

6

7

8

9

10

133. DOVE POSE

1. ILIOPSOAS
2. TENSOR FASCIA LATA
3. RECTUS ABDOMINIS
4. LATISSIMUS DORSI
5. QUADRICEPS
6. PECTORALIS MAJOR
7. HAMSTRINGS
8. GLUTEUS MAXIMUS
9. ERECTOR SPINAE
10. TRICEPS BRACHII

134. BOUND ANGLE HEADSTAND POSE

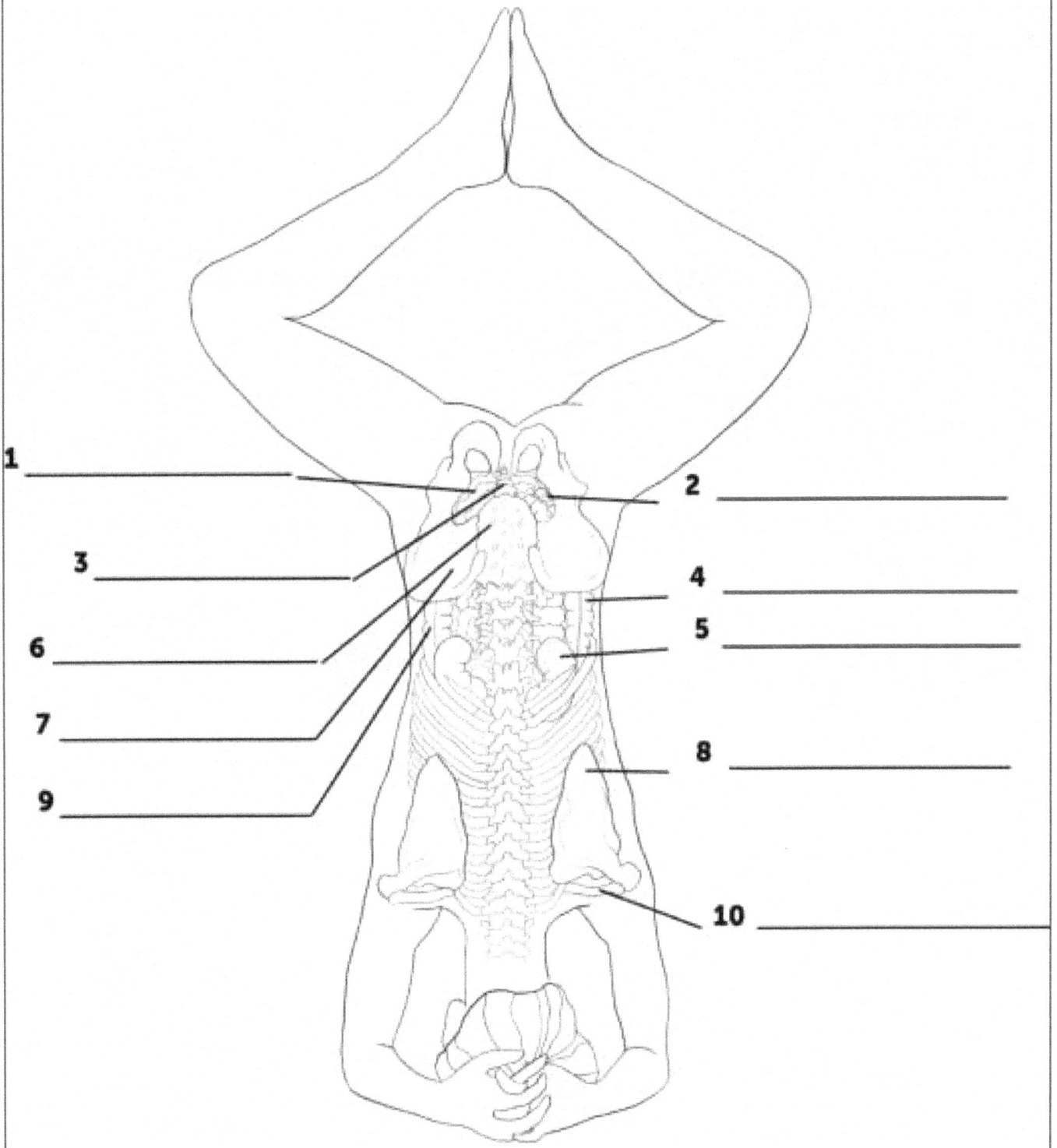

1 _____

2 _____

3 _____

4 _____

5 _____

6 _____

7 _____

8 _____

9 _____

10 _____

134. BOUND ANGLE HEADSTAND POSE

1. COILS OF SMALL INTESTINE
2. SIGMOID COLON
3. COCCYX
4. DESCENDING COLON
5. KIDNEY
6. SACRUM
7. PELVIS
8. SCAPULA
9. ASCENDING COLON
10. COLLARBONE

135. VISVAMITRASANA II

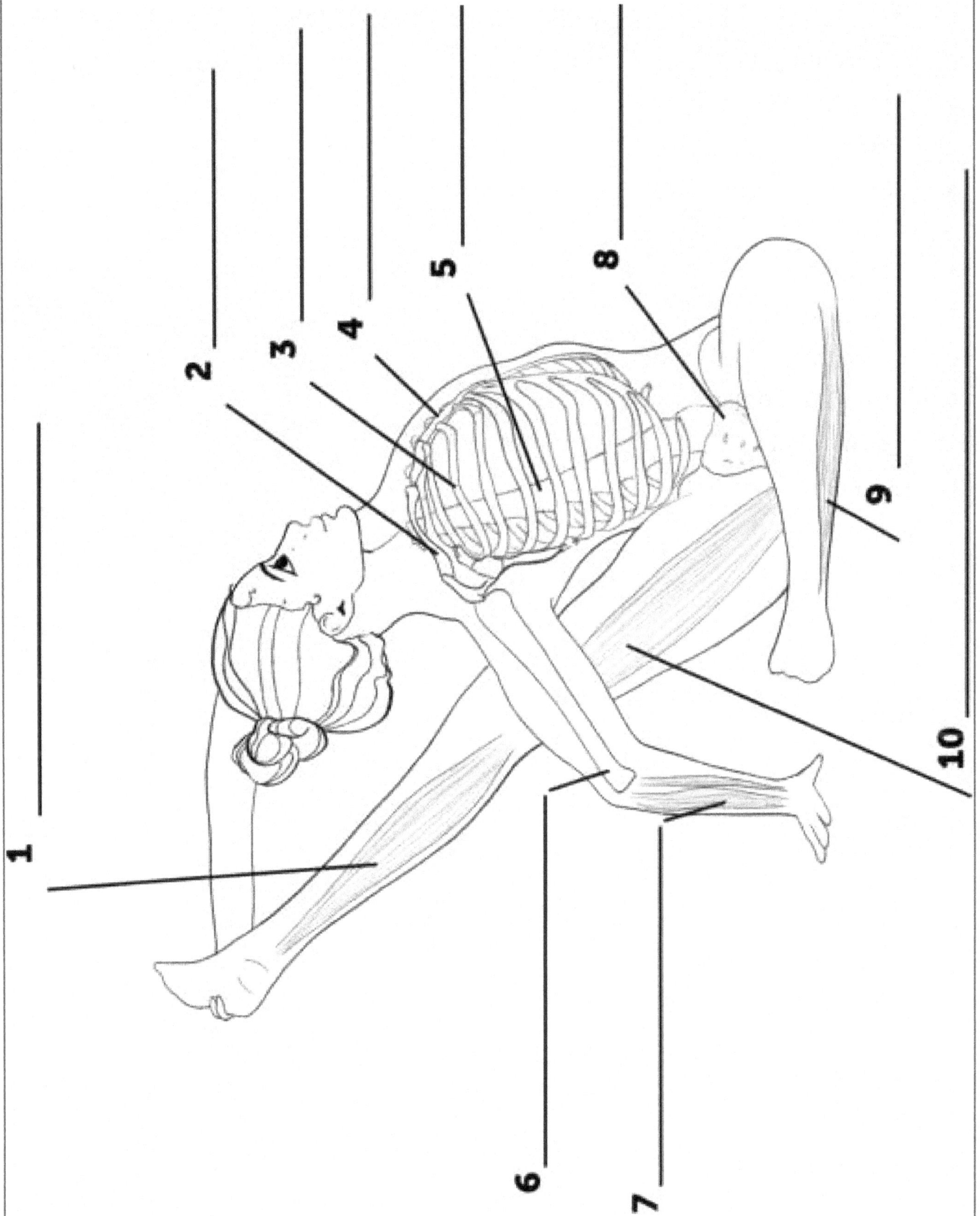

1
2
3
4
5
6
7
8
9
10

135. VISVAMITRASANA II

1. GASTROCNEMIUS
2. COLLARBONE
3. RIBS
4. STERNUM
5. SPINE
6. HUMERUS
7. PRONATORS
8. SACRUM
9. TIBIALIS ANTERIOR
10. HAMSTRINGS

136. LOTUS IN SHOULDER STAND POSE

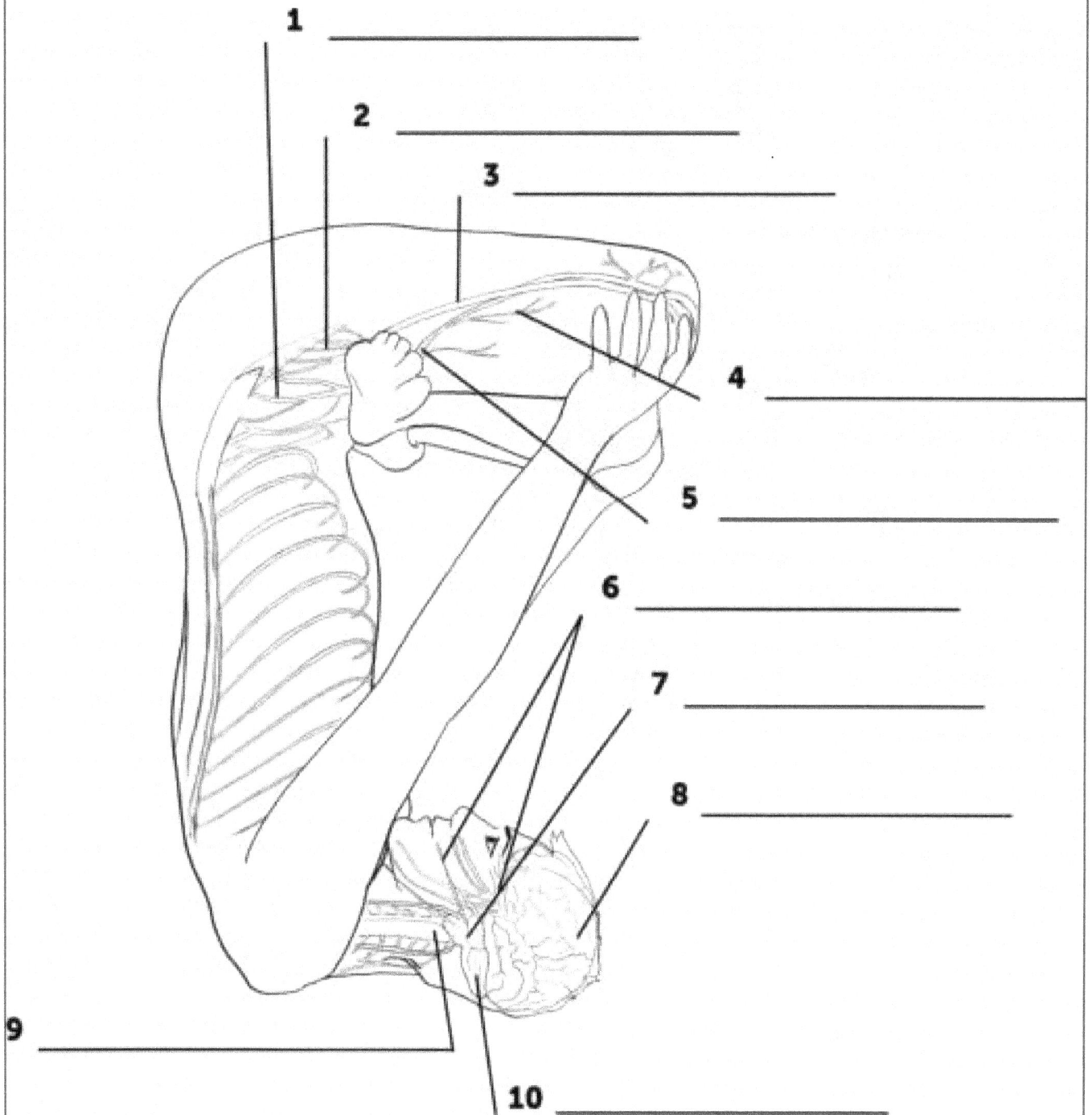

1 _____

2 _____

3 _____

4 _____

5 _____

6 _____

7 _____

8 _____

9 _____

10 _____

136. LOTUS IN SHOULDER STAND POSE

1. LUMBAR PLEXUS
2. SACRAL PLEXUS
3. SCIATIC
4. MUSCULAR BRANCHES OF FEMORAL
5. FEMORAL
6. CRANIAL NERVES
7. BRAINSTEM
8. CEREBRUM
9. SPINAL CORD
10. CEREBELLUM

137. ONE LEGGED WHEEL POSE

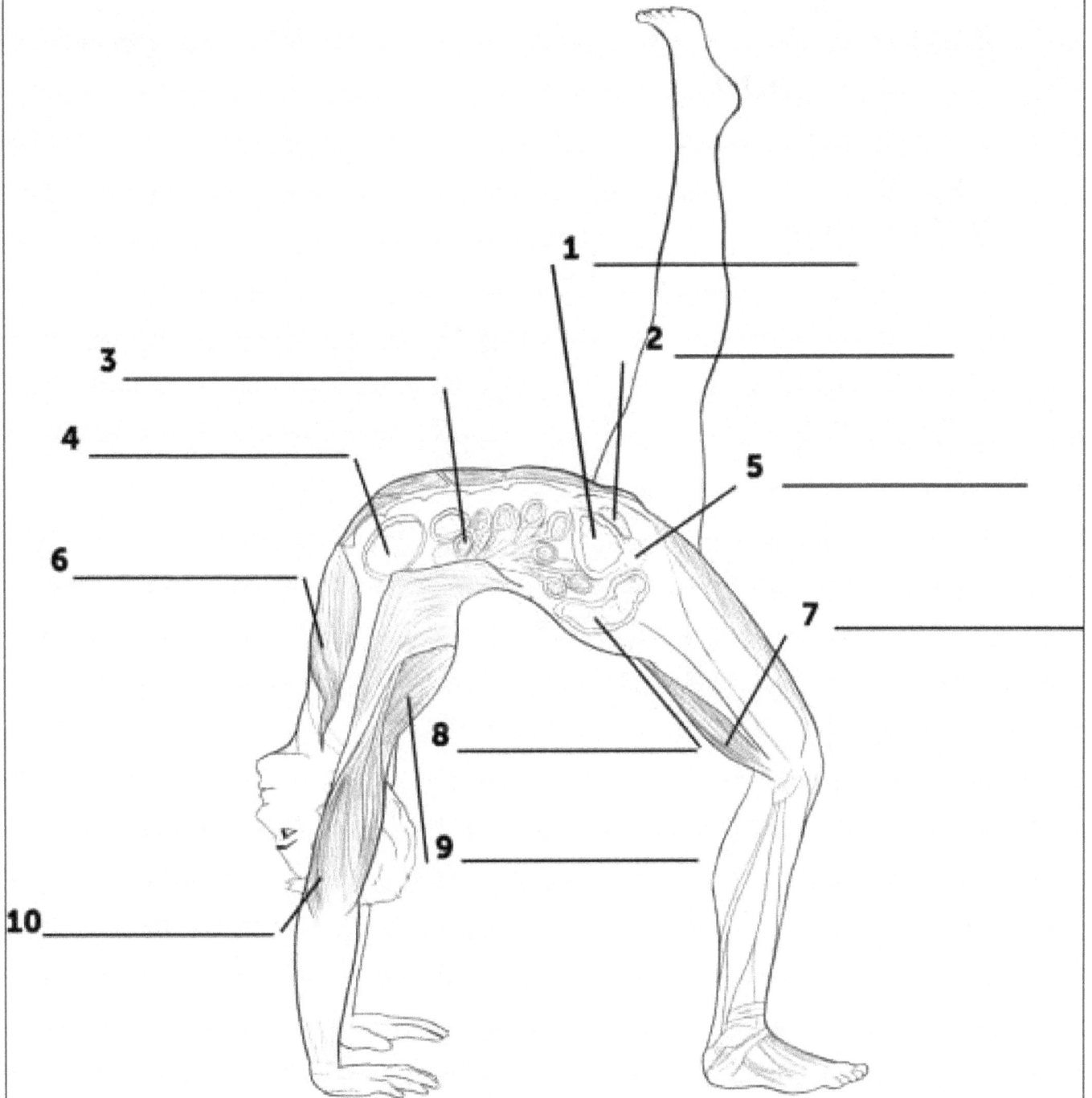

1 _____

2 _____

3 _____

4 _____

5 _____

6 _____

7 _____

8 _____

9 _____

10 _____

137. ONE LEGGED WHEEL POSE

1. URINARY BLADDER
2. PUBIC BONE
3. COILS OF SMALL INTESTINE
4. STOMACH
5. PROSTATE
6. PECTORALIS MAJOR
7. HAMSTRINGS
8. RECTUM
9. ERECTOR SPINAE
10. TRICEPS BRACHII

138. ONE LEGGED HEADSTAND

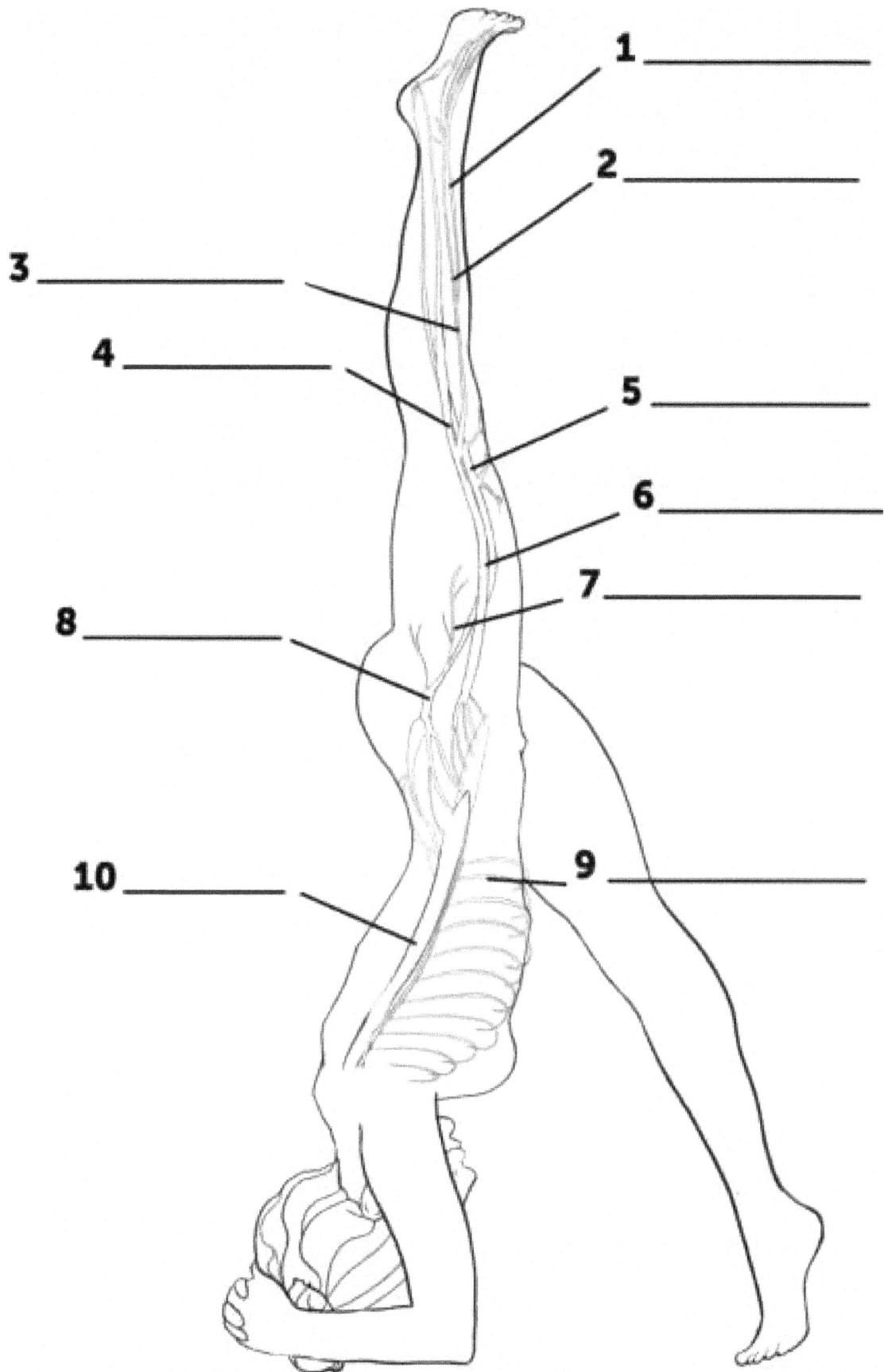

1 _____

2 _____

3 _____

4 _____

5 _____

6 _____

7 _____

8 _____

9 _____

10 _____

138. ONE LEGGED HEADSTAND

1. SUPERFICIAL PERONEAL

2. DEEP PERONEAL

3. COMMON PERONEAL

4. TIBIAL

5. SAPHENOUS

6. SCIATIC

7. MUSCULAR BRANCHES OF FEMORAL

8. FEMORAL

9. INTERCOSTALS

10. SPINAL CORD

139. SUPTA VISVAMITRASANA

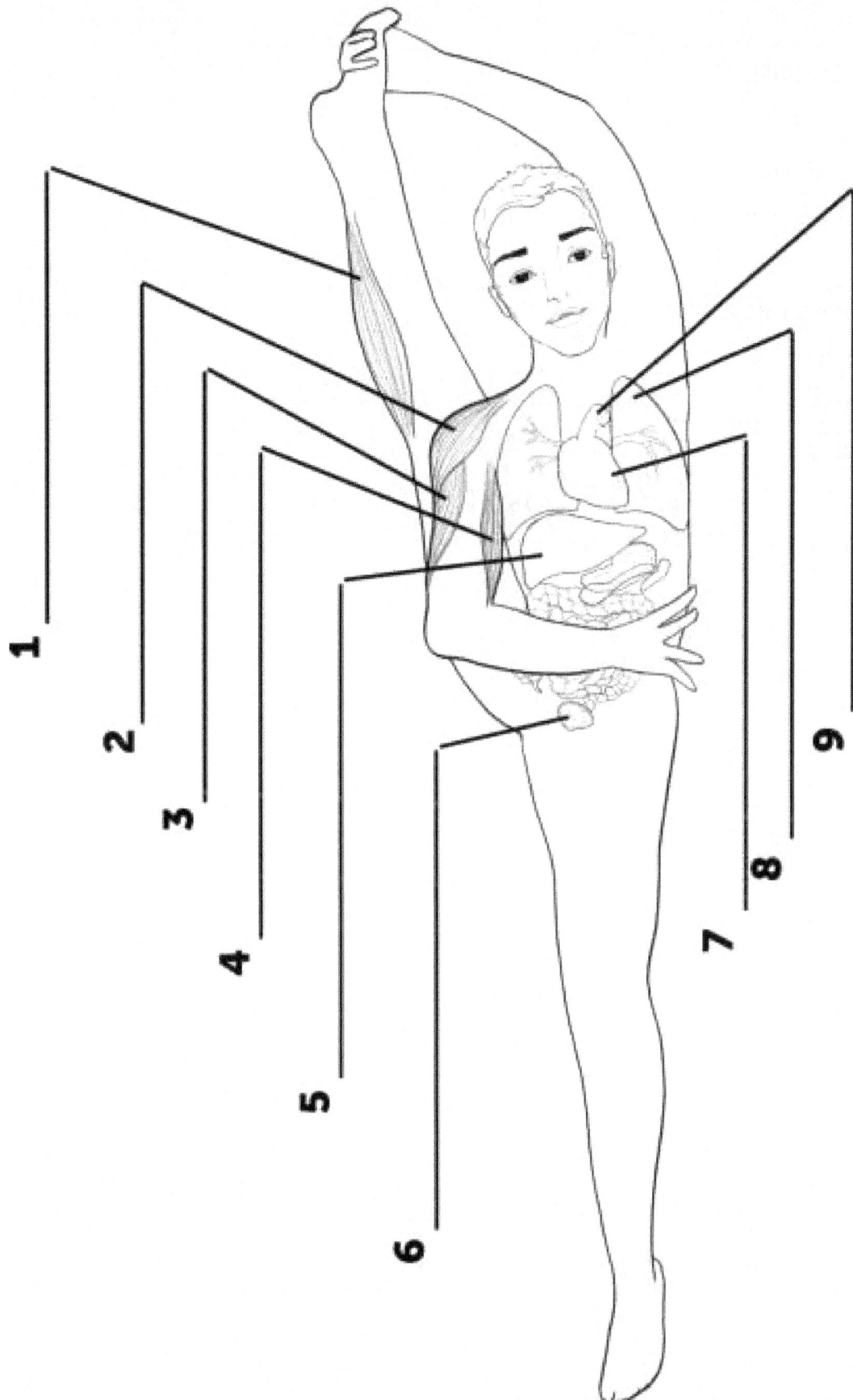

1
2
3
4
5
6
7
8
9

139. SUPTA VISVAMITRASANA

1. GASTROCNEMIUS
2. DELTOID
3. TRICEPS BRACHII
4. BICEPS BRACHII
5. LIVER
6. URINARY BLADDER
7. HEART
8. LUNGS
9. AORTA

140. UPWARD FACING FORWARD BEND POSE

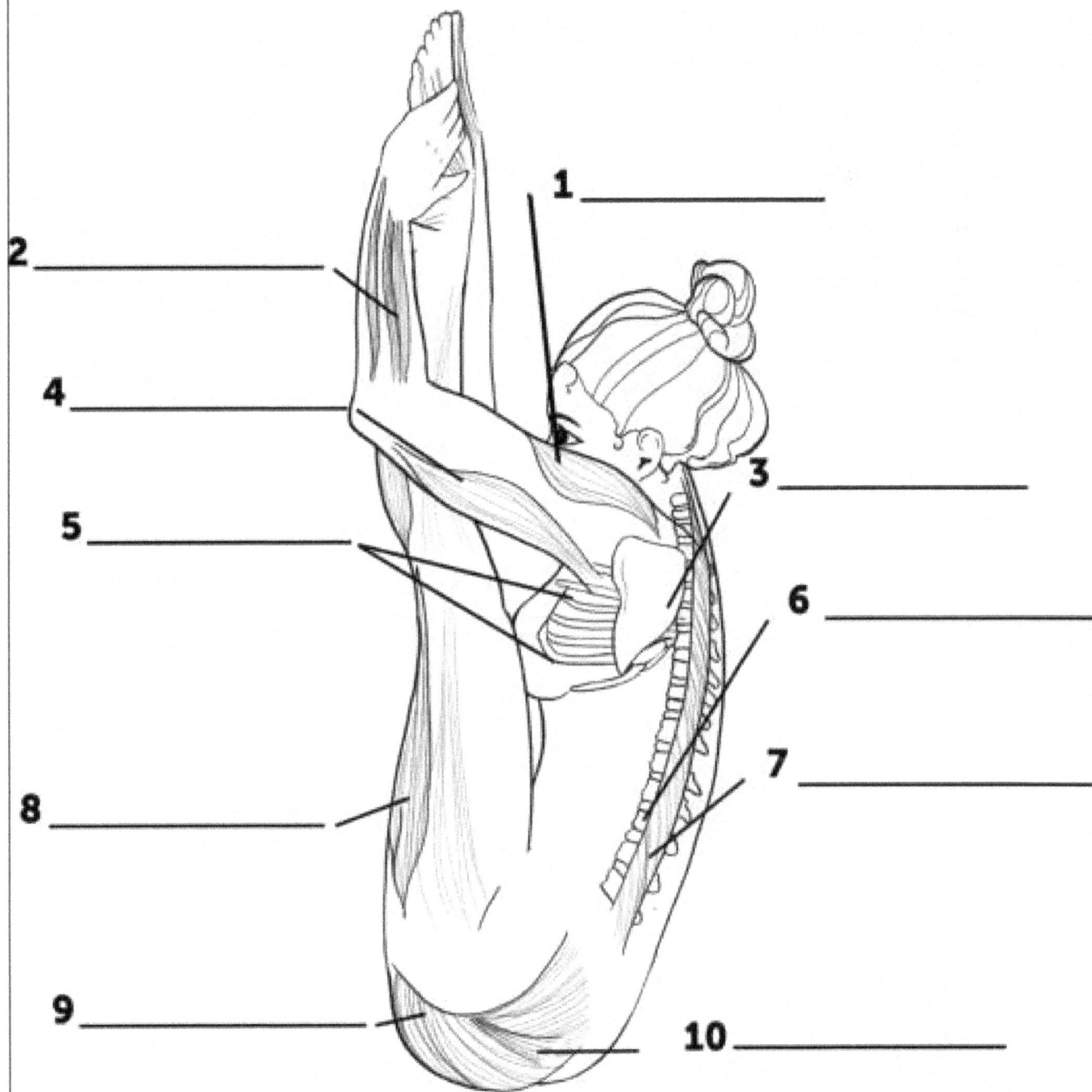

1 _____

2 _____

4 _____

5 _____

8 _____

9 _____

3 _____

6 _____

7 _____

10 _____

140. UPWARD FACING FORWARD BEND POSE

1. DELTOID
2. PRONATORS
3. SCAPULA
4. TRICEPS BRACHII
5. RIBS
6. SPINE
7. SPINAL MUSCLES
8. HAMSTRINGS
9. GLUTEUS MAXIMUS
10. PIRIFORMIS

141. UPWARD FACING WIDE-ANGLE SEATED POSE

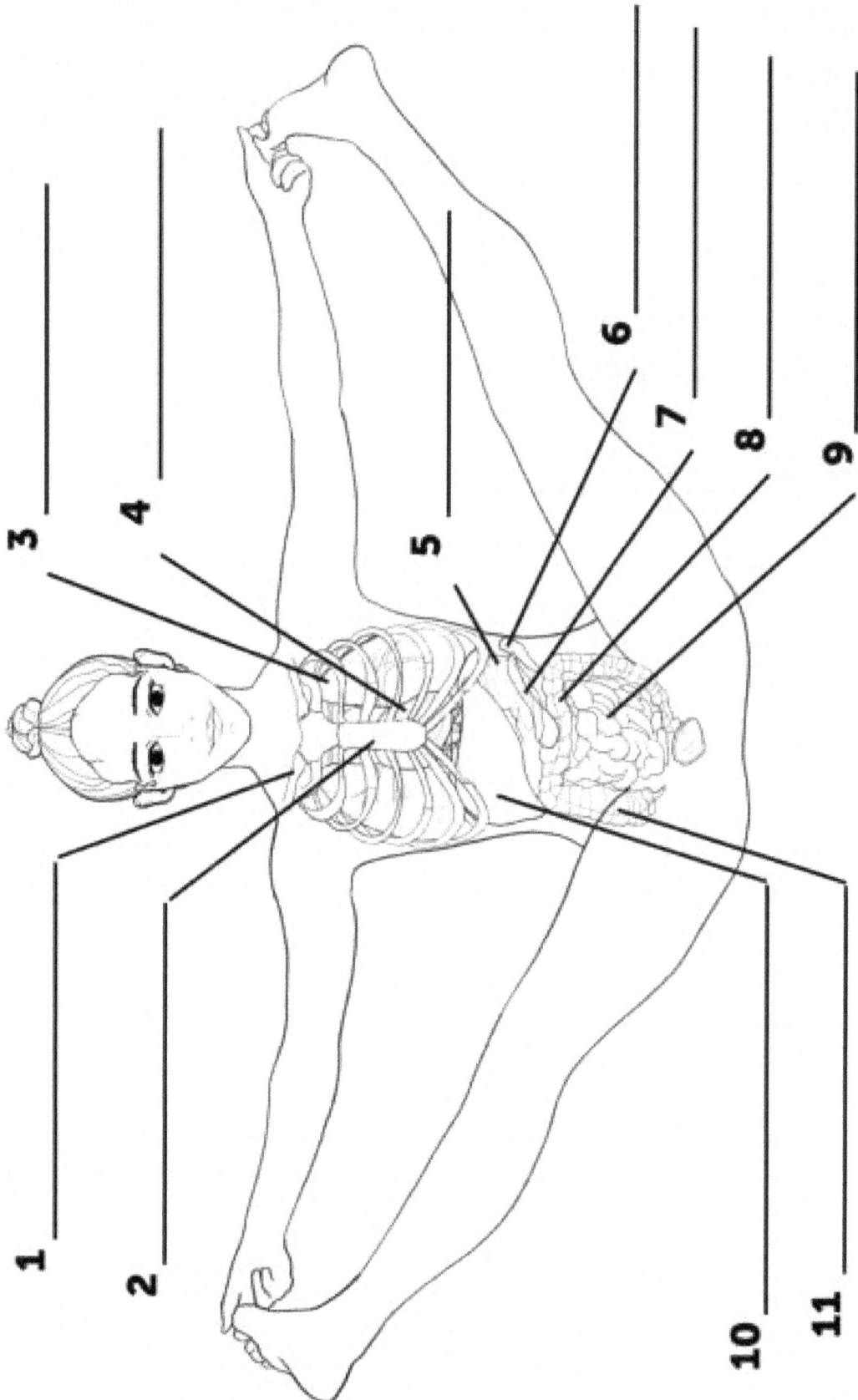

1

2

3

4

5

6

7

8

9

10

11

141. UPWARD FACING WIDE-ANGLE SEATED POSE

1. COLLARBONE
2. STERNUM
3. LUNGS
4. HEART
5. STOMACH
6. SPLEEN
7. PANCREAS
8. TRANSVERSE COLON
9. COILS OF SMALL INTESTINE
10. LIVER
11. ASCENDING COLON

142. VISVAMITRASANA

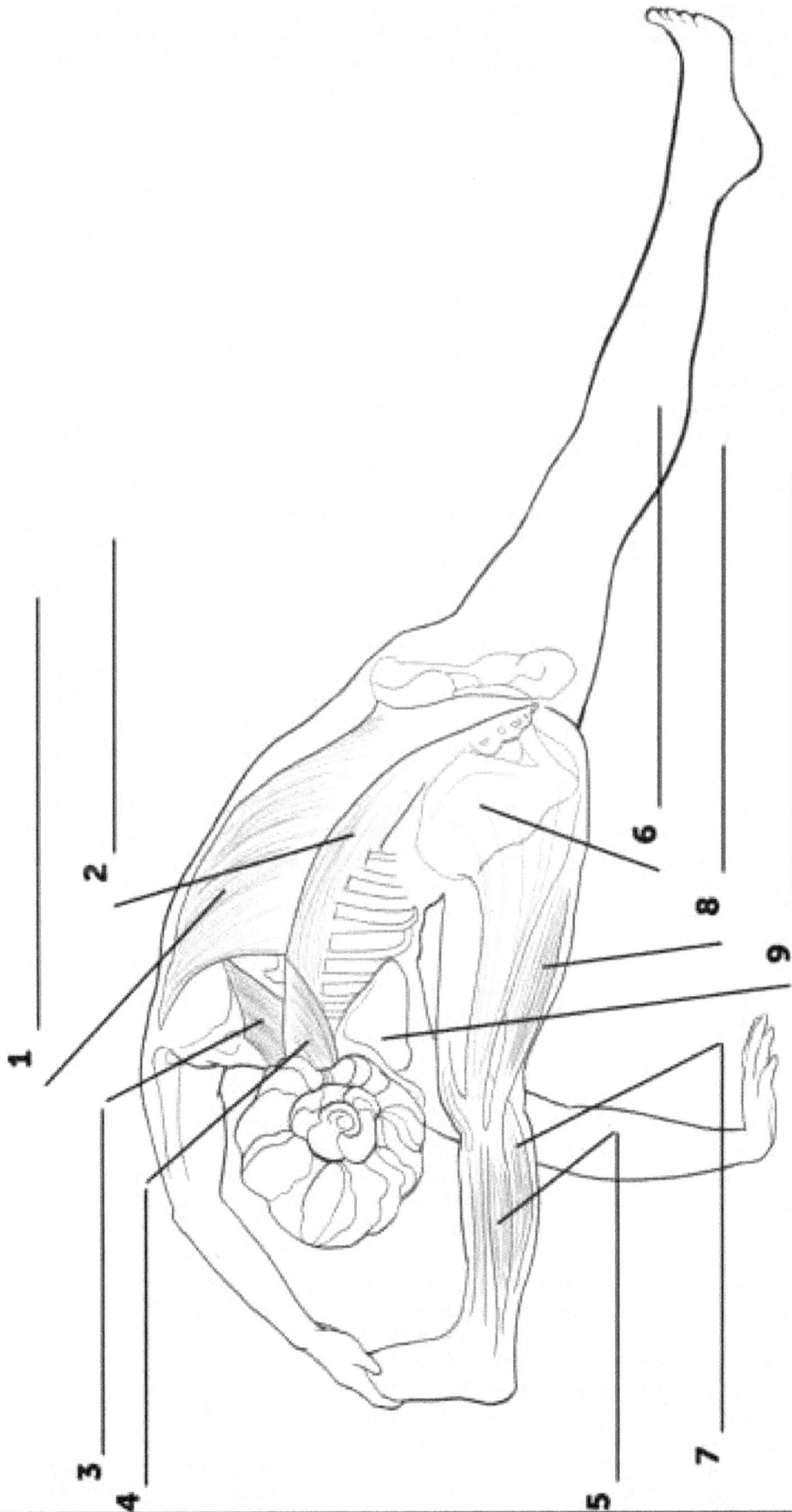

142. VISVAMITRASANA

1. LATISSIMUS DORSI
2. ERECTOR SPINAE
3. RHOMBOIDS
4. TRAPEZIUS
5. SOLEUS
6. PELVIS
7. GASTROCNEMIUS
8. HAMSTRINGS
9. SCAPULA

143. BOUND SKANDASANA

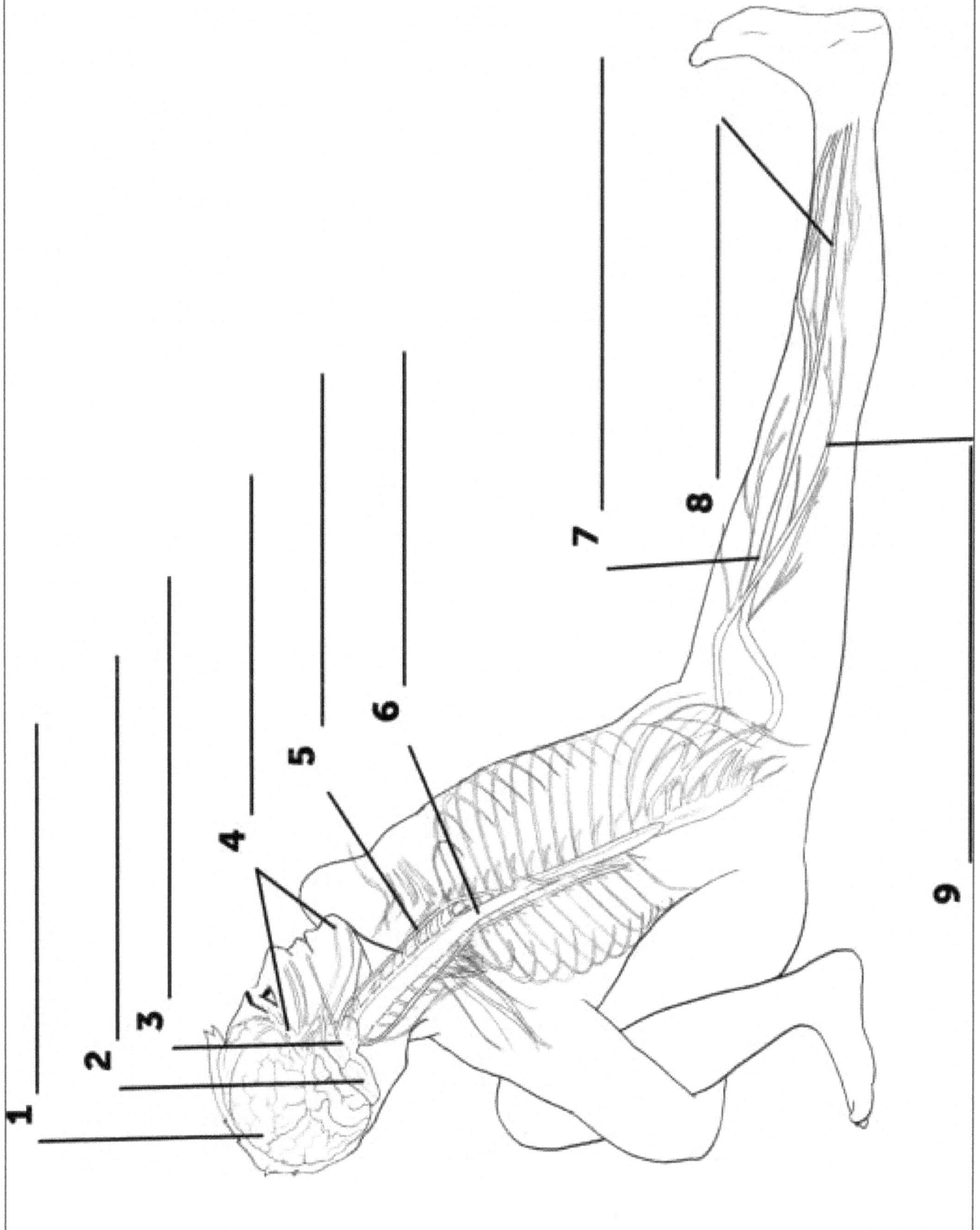

143. BOUND SKANDASANA

1. CEREBRUM
2. CEREBELLUM
3. BRAINSTEM
4. CRANIAL NERVES
5. VAGUS
6. SPINAL CORD
7. SCIATIC
8. TIBIAL
9. SAPHENOUS

144. DEVOTIONAL WARRIOR POSE

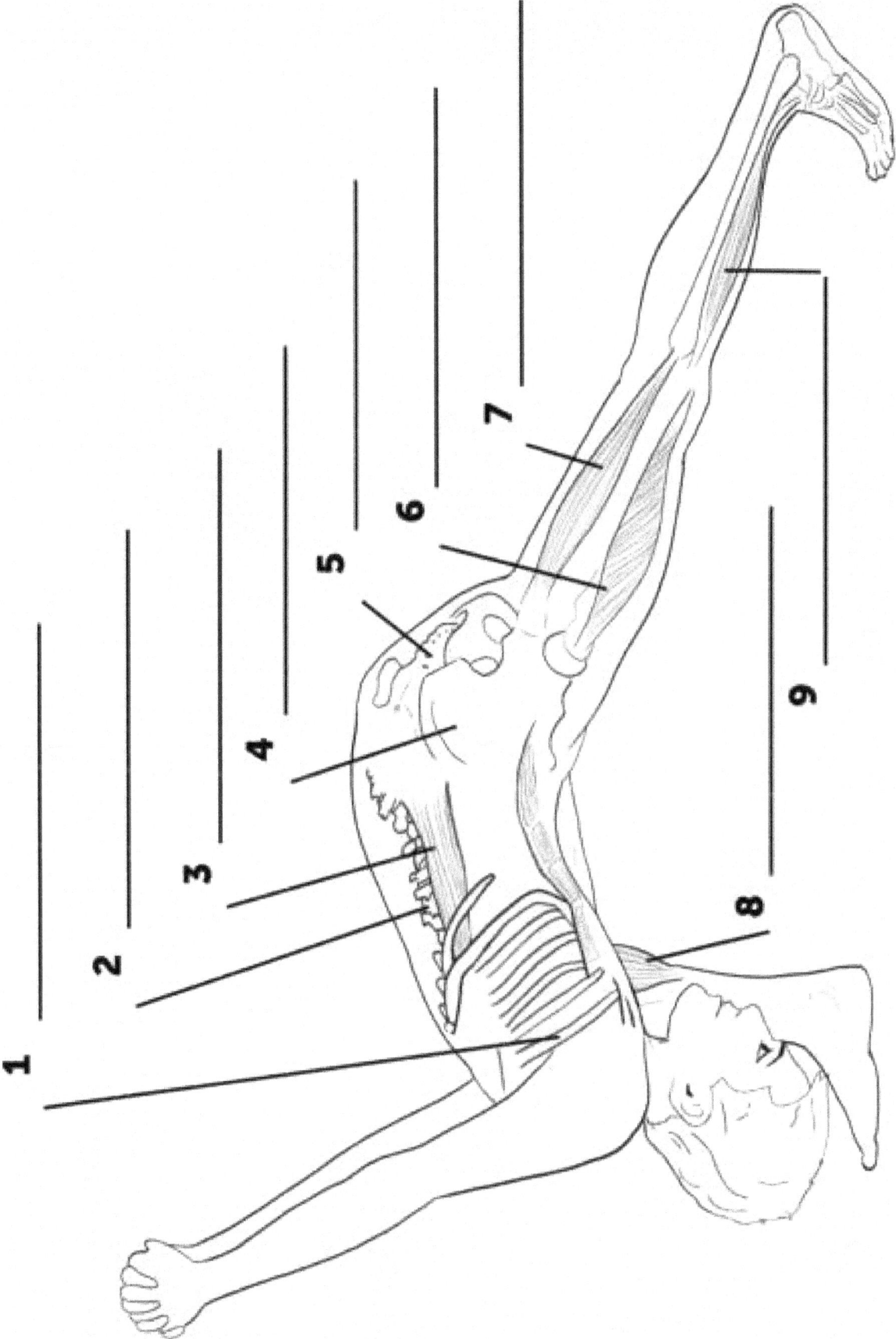

144. DEVOTIONAL WARRIOR POSE

1. RIBS
2. SPINE
3. ERECTOR SPINAE
4. PELVIS
5. SACRUM
6. QUADRICEPS
7. HAMSTRINGS
8. GASTROCNEMIUS
9. TIBIALIS ANTERIOR

145. BOUND LIZARD POSE

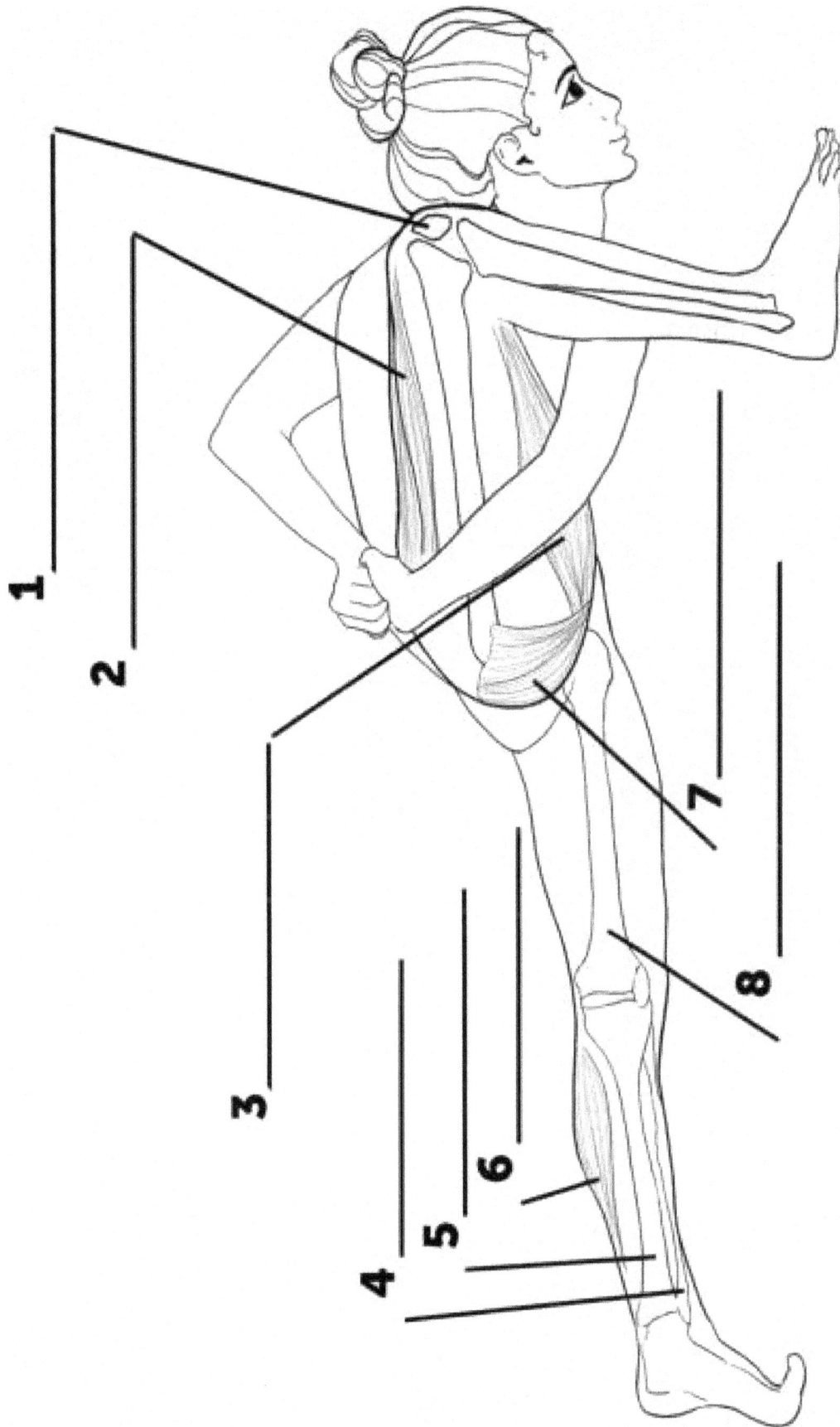

145. BOUND LIZARD POSE

1. PATELLA
2. QUADRICEPS
3. HAMSTRINGS
4. FIBULA
5. TIBIA
6. GASTROCNEMIUS
7. GLUTEUS MAXIMUS
8. FEMUR

146. STANDING SPLIT

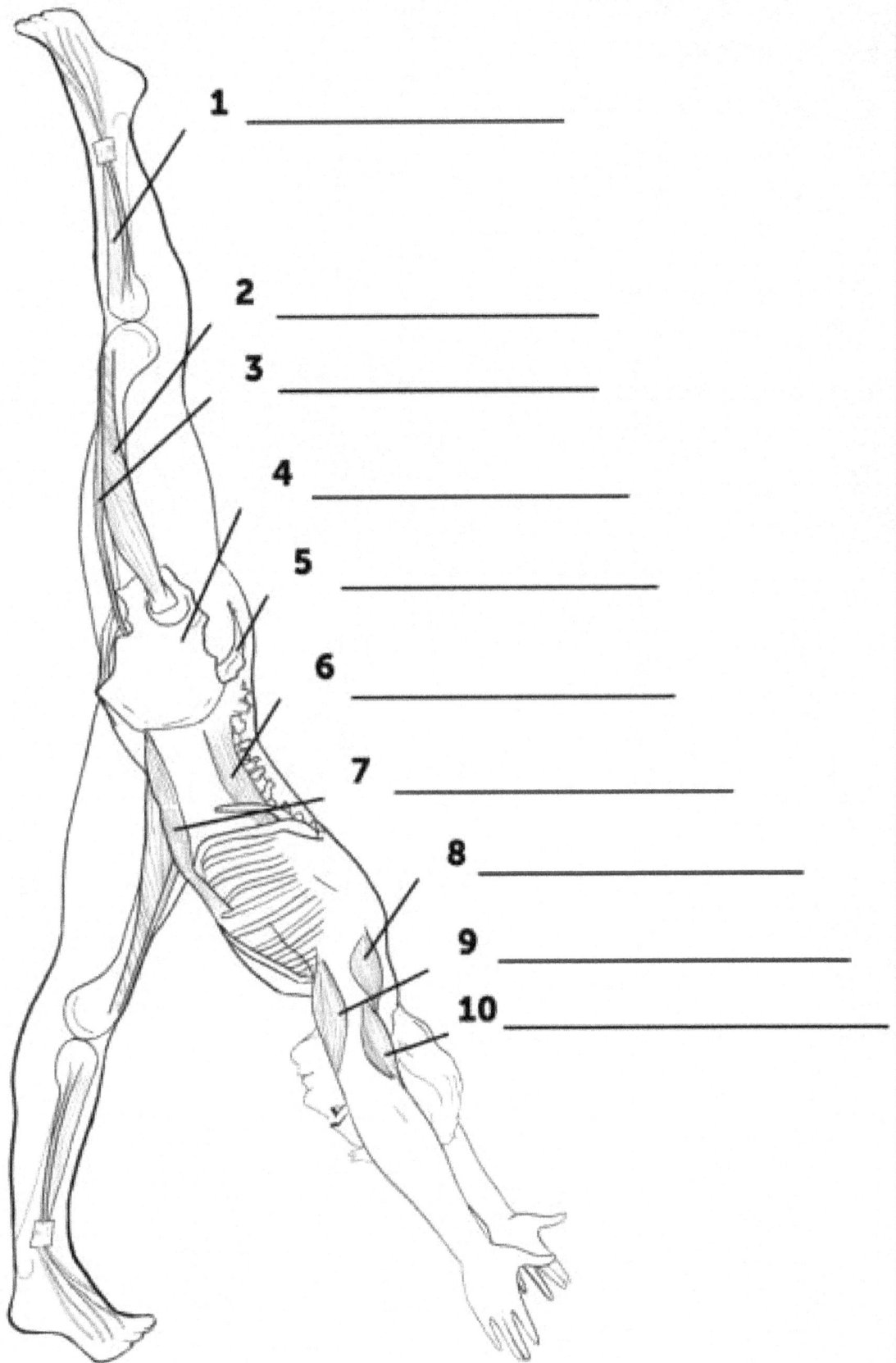

1 _____

2 _____

3 _____

4 _____

5 _____

6 _____

7 _____

8 _____

9 _____

10 _____

146. STANDING SPLIT

1. TIBIALIS ANTERIOR
2. RECTUS FEMORIS
3. SARTORIUS
4. PELVIS
5. SACRUM
6. ERECTOR SPINAE
7. RECTUS ABDOMINIS
8. DELTOID
9. BICEPS BRACHII
10. TRICEPS BRACHII

147. BOUND WARRIOR III

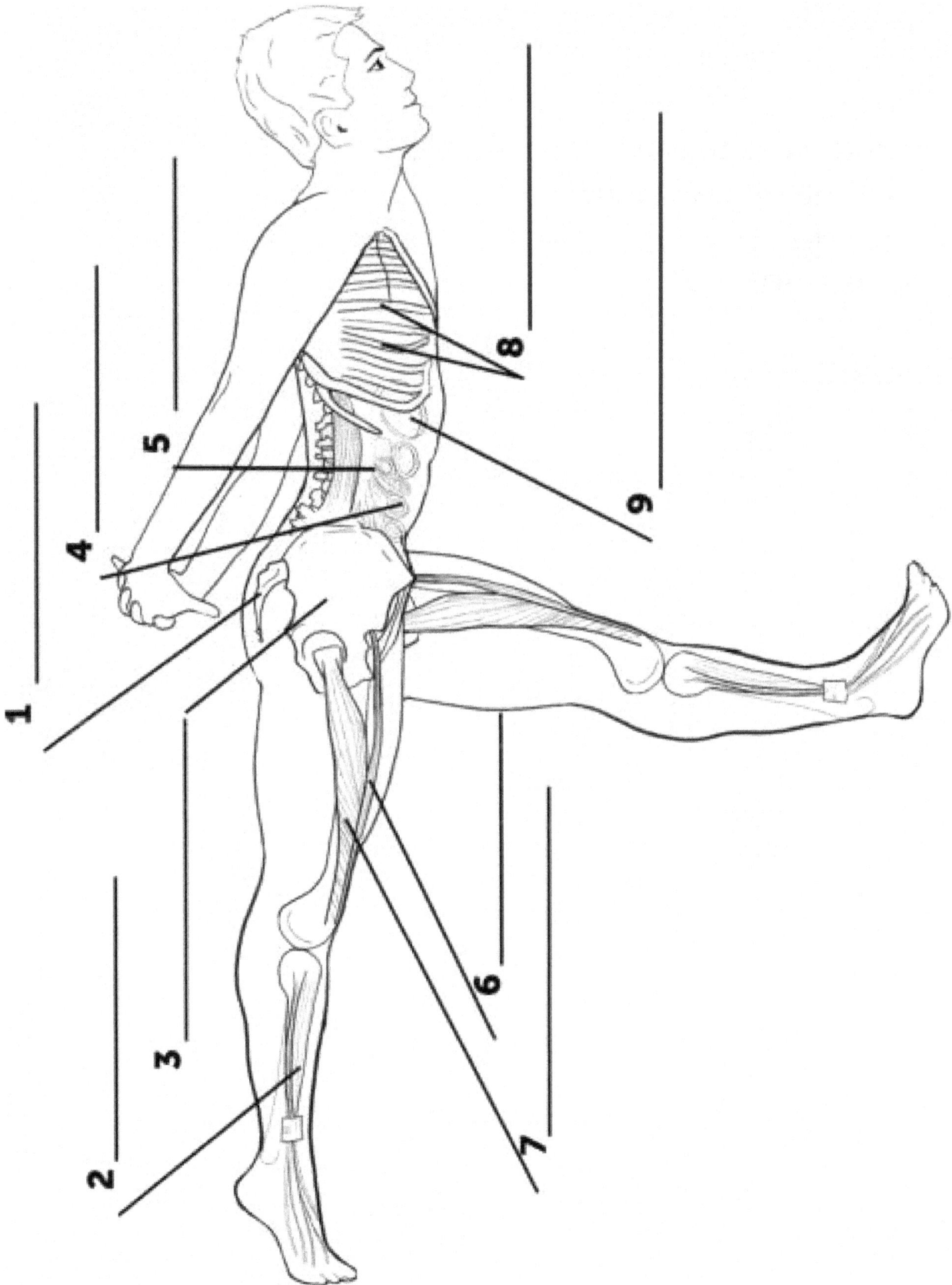

147. BOUND WARRIOR III

1. SACRUM
2. TIBIALIS ANTERIOR
3. PELVIS
4. COILS OF SMALL INTESTINE
5. MESENTERY OF SMALL INTESTINE
6. SARTORIUS
7. RECTUS FEMORIS
8. RIBS
9. STOMACH

148. BOUND FORWARD FOLD

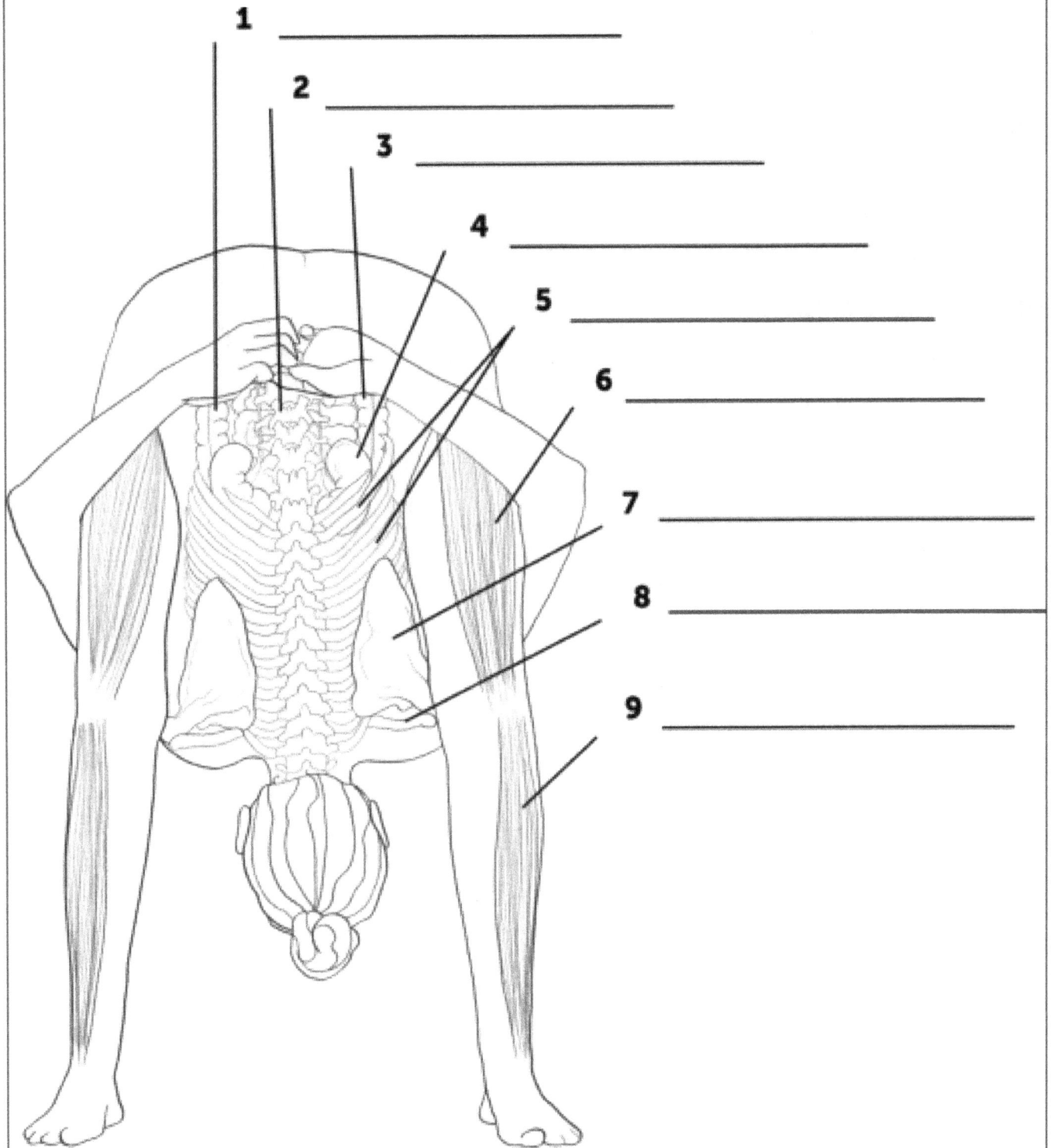

1 _____

2 _____

3 _____

4 _____

5 _____

6 _____

7 _____

8 _____

9 _____

148. BOUND FORWARD FOLD

1. ASCENDING COLON
2. SPINE
3. DESCENDING COLON
4. KIDNEY
5. RIBS
6. QUADRICEPS
7. SCAPULA
8. COLLARBONE
9. TIBIALIS ANTERIOR

149. RAG DOLL POSE

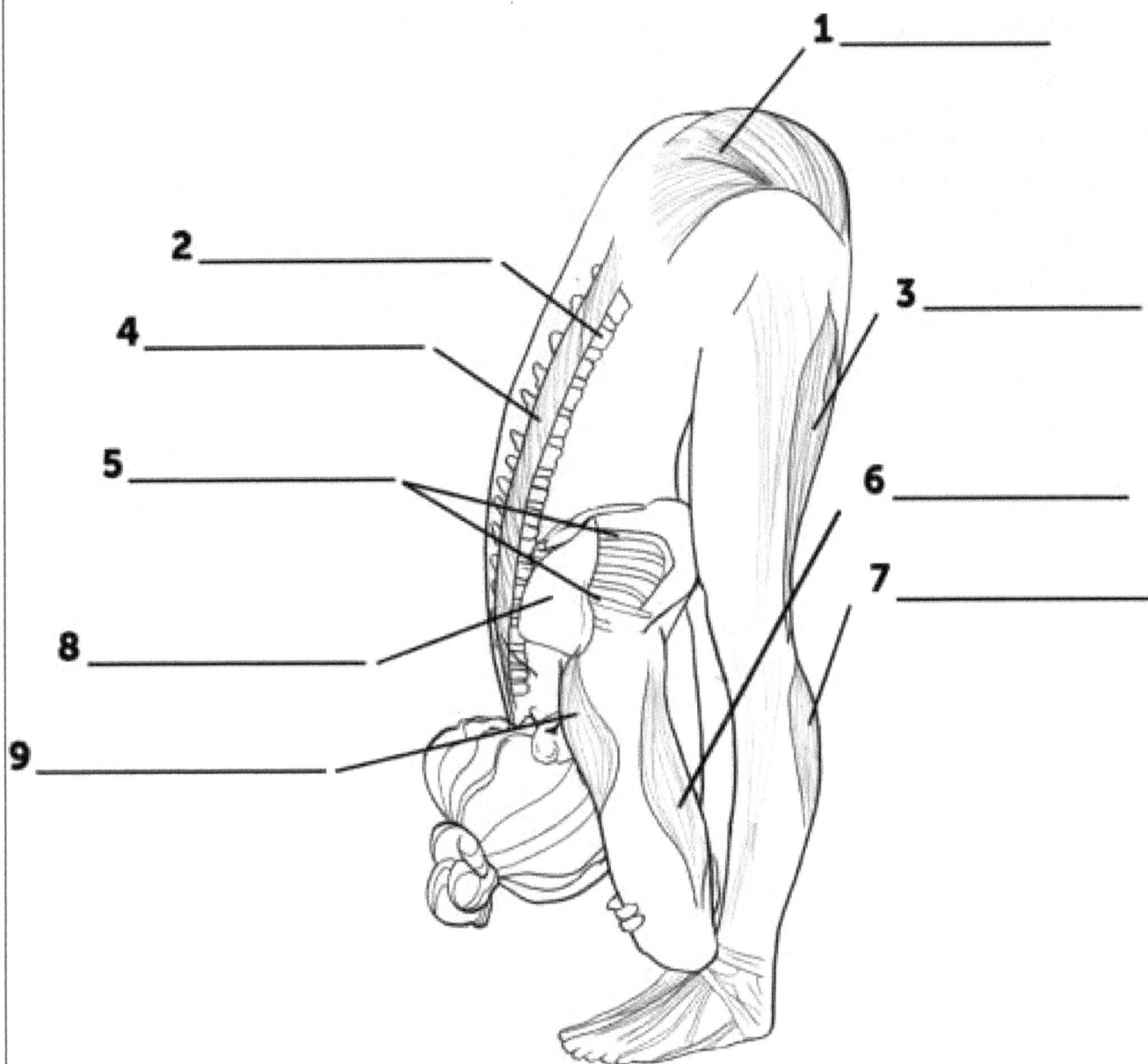

1 _____

2 _____

3 _____

4 _____

5 _____

6 _____

7 _____

8 _____

9 _____

149. RAG DOLL POSE

1. PIRIFORMIS
2. SPINE
3. HAMSTRINGS
4. SPINAL MUSCLES
5. RIBS
6. TRICEPS BRACHII
7. GASTROCNEMIUS
8. SCAPULA
9. DELTOID

150. RESTED HALF PIGEON POSE

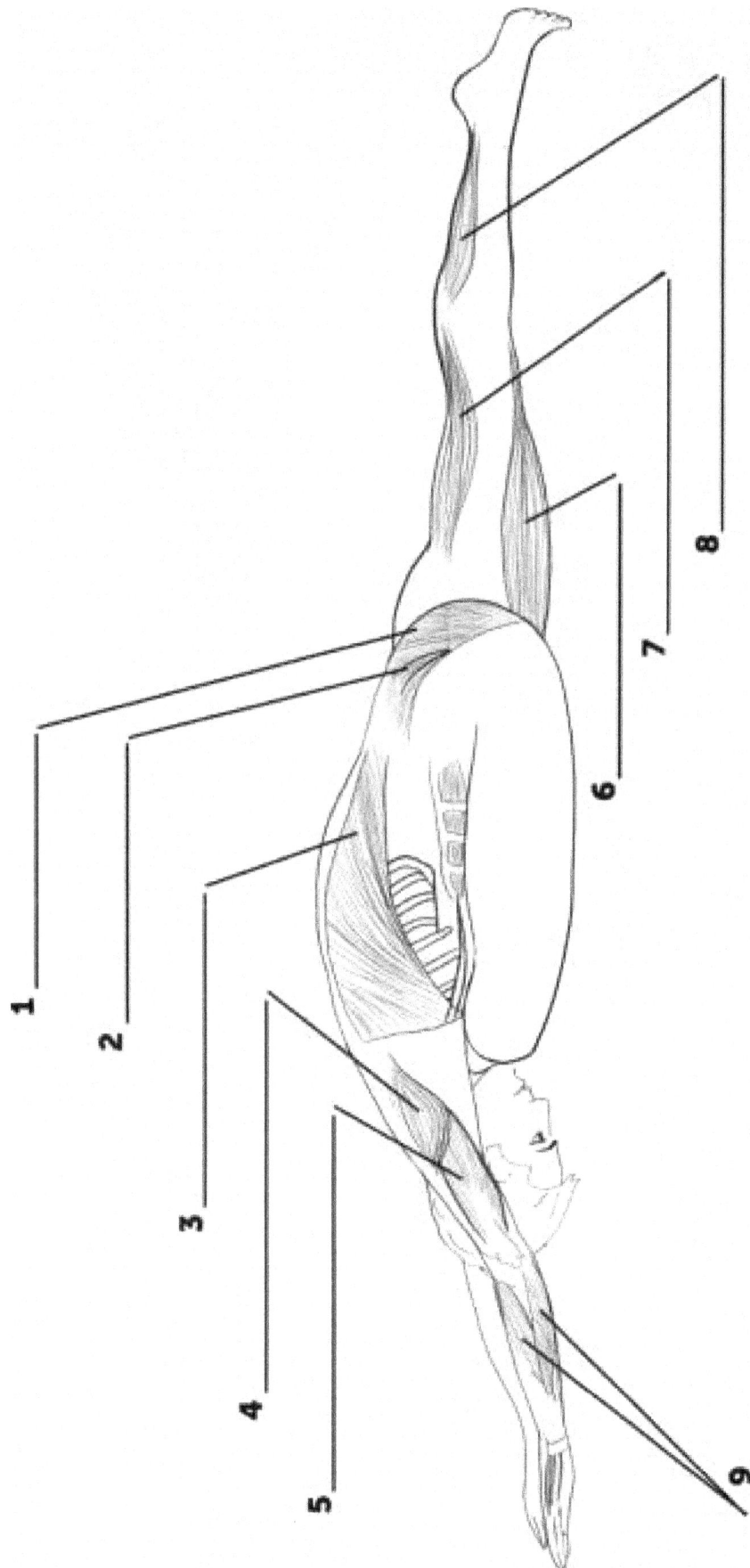

1

2

3

4

5

6

7

8

9

150. RESTED HALF PIGEON POSE

1. GLUTEUS MAXIMUS
2. PIRIFORMIS
3. LATISSIMUS DORSI
4. DELTOID
5. TRICEPS BRACHII
6. QUADRICEPS
7. HAMSTRINGS
8. GASTROCNEMIUS
9. PRONATORS

151. ONE LEGGED REVERSE TABLE

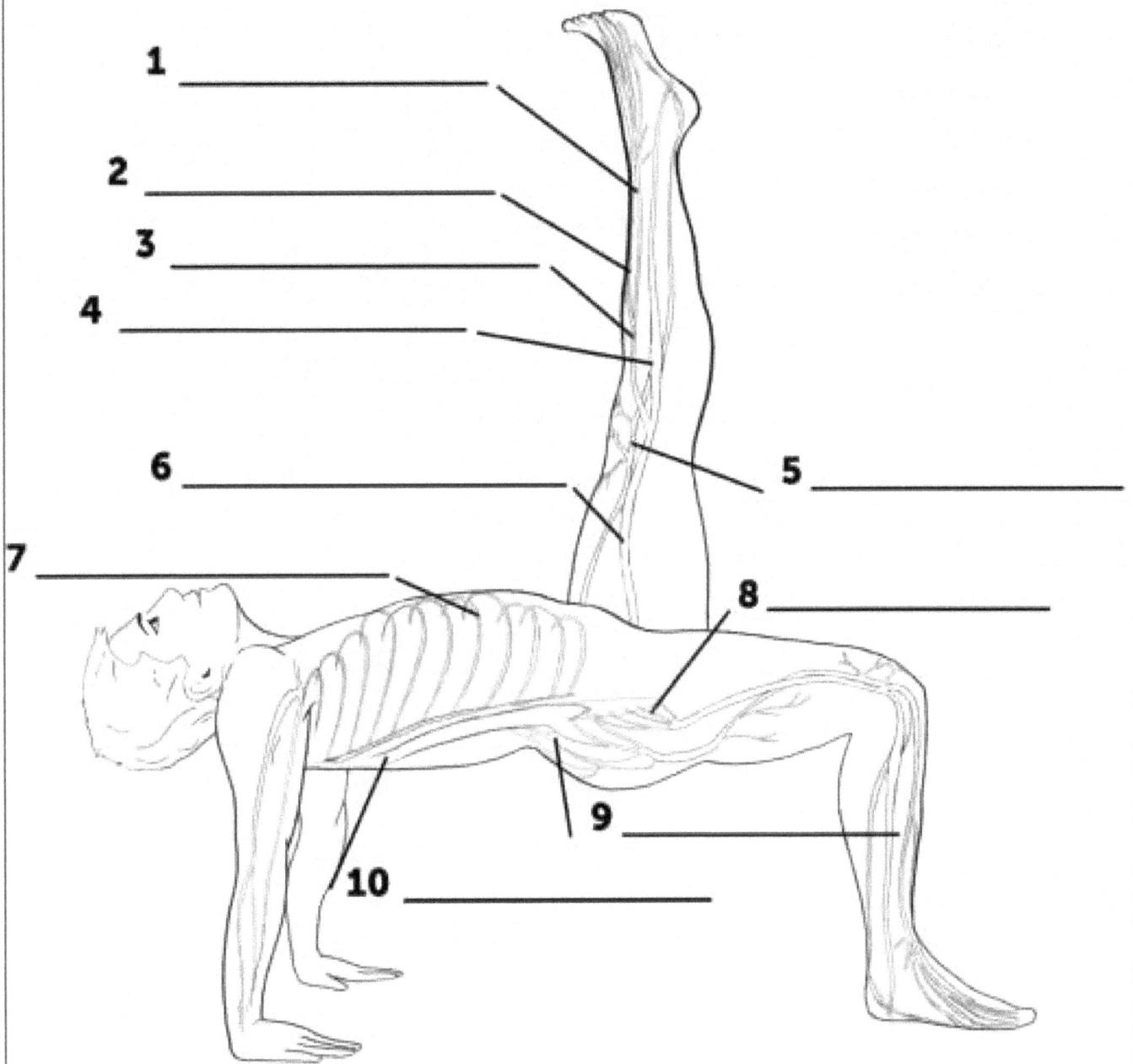

1 _____

2 _____

3 _____

4 _____

6 _____

5 _____

7 _____

8 _____

9 _____

10 _____

151. ONE LEGGED REVERSE TABLE

1. DEEP PERONEAL
2. SUPERFICIAL PERONEAL
3. COMMON PERONEAL
4. TIBIAL
5. SAPHENOUS
6. SCIATIC
7. INTERCOSTALS
8. SACRAL PLEXUS
9. LUMBAR PLEXUS
10. SPINAL CORD

152. ONE LEGGED CROW II

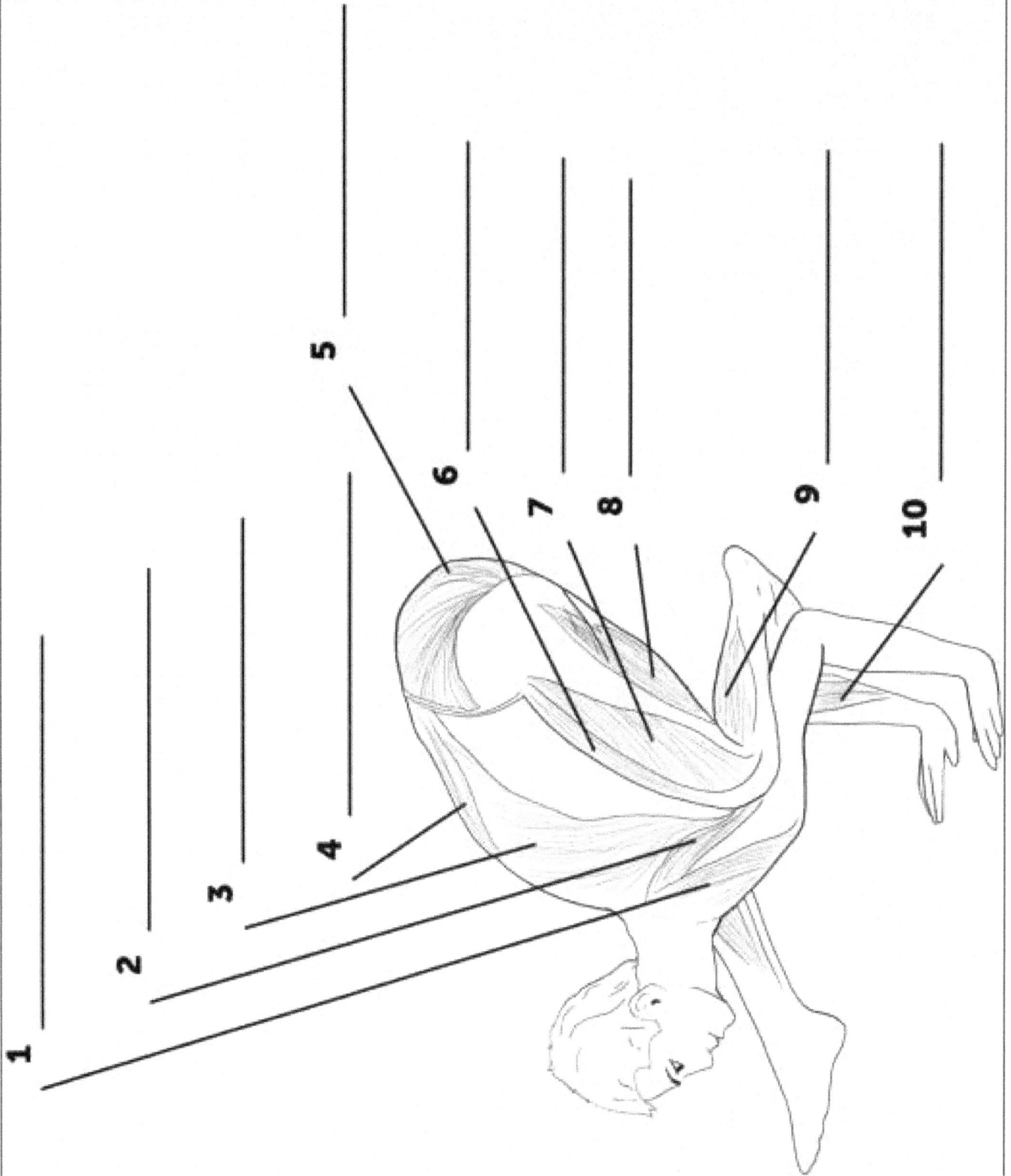

1

2

3

4

5

6

7

8

9

10

152. ONE LEGGED CROW II

1. DELTOID
2. TRICEPS BRACHII
3. LATISSIMUS DORSI
4. ERECTOR SPINAE
5. GLUTEUS MAXIMUS
6. RECTUS FEMORIS
7. VASTUS LATERALIS
8. HAMSTRINGS
9. GASTROCNEMIUS
10. PRONATORS

153. DRAGONFLY

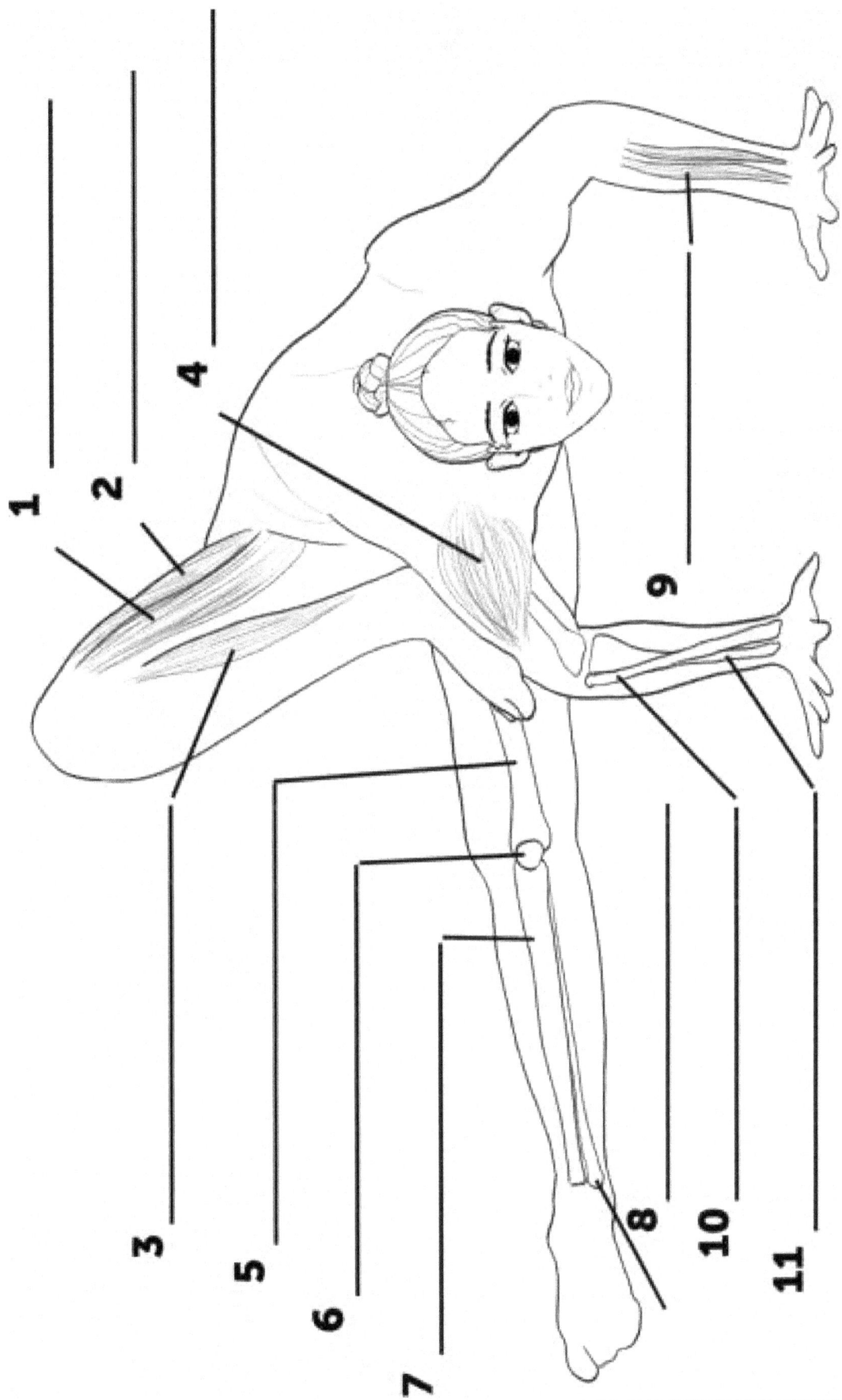

1

2

3

4

5

6

7

8

9

10

11

153. DRAGONFLY

1. VASTUS LATERALIS
2. RECTUS FEMORIS
3. GASTROCNEMIUS
4. DELTOID
5. FEMUR
6. PATELLA
7. TIBIA
8. FIBULA
9. PRONATORS
10. RADIUS
11. ULNA

154. ONE HANDED TREE POSE

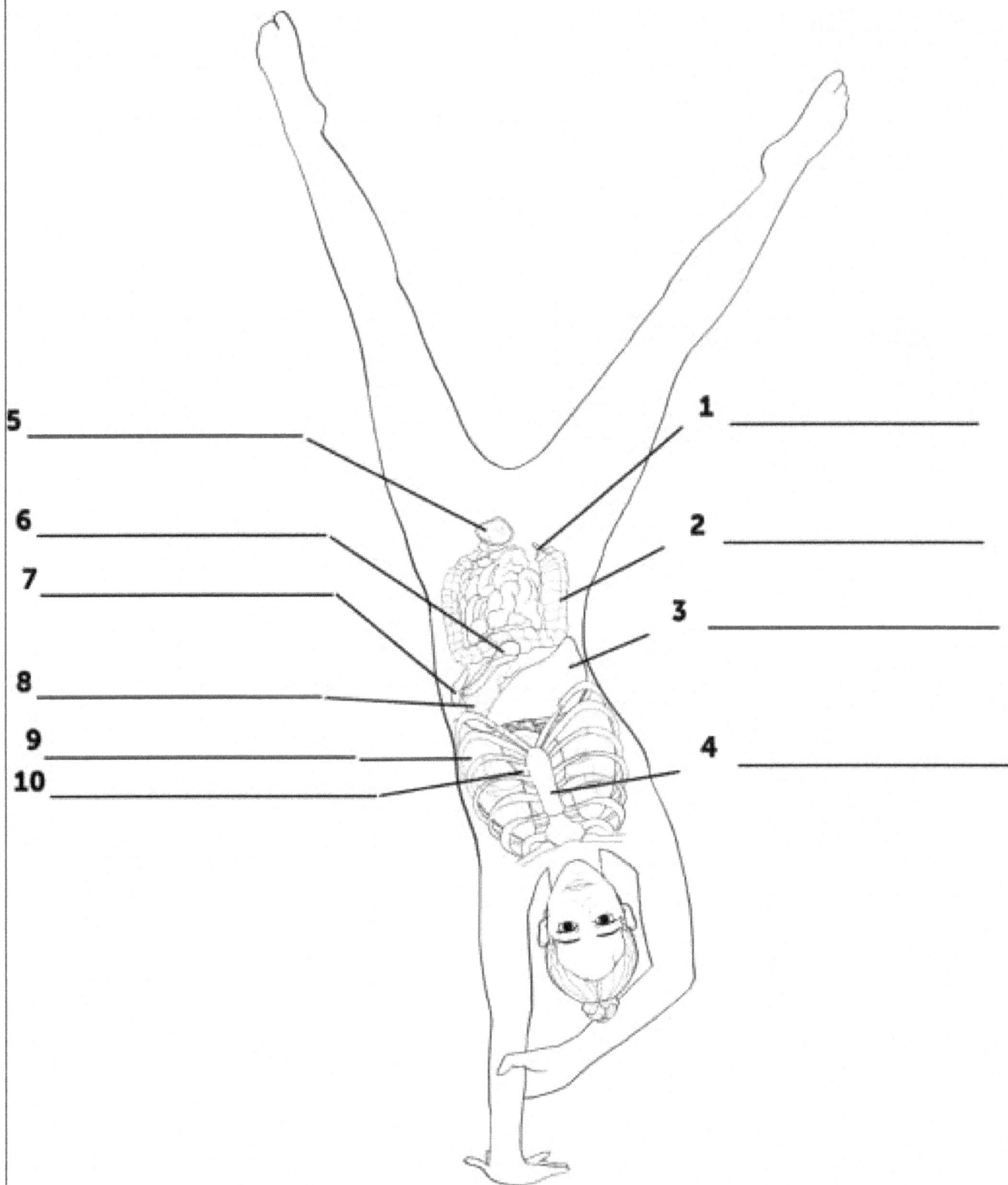

5 _____

6 _____

7 _____

8 _____

9 _____

10 _____

1 _____

2 _____

3 _____

4 _____

154. ONE HANDED TREE POSE

1. APPENDIX
2. ASCENDING COLON
3. LIVER
4. STERNUM
5. URINARY BLADDER
6. PANCREAS
7. SPLEEN
8. STOMACH
9. LUNGS
10. HEART

155. KING COBRA POSE

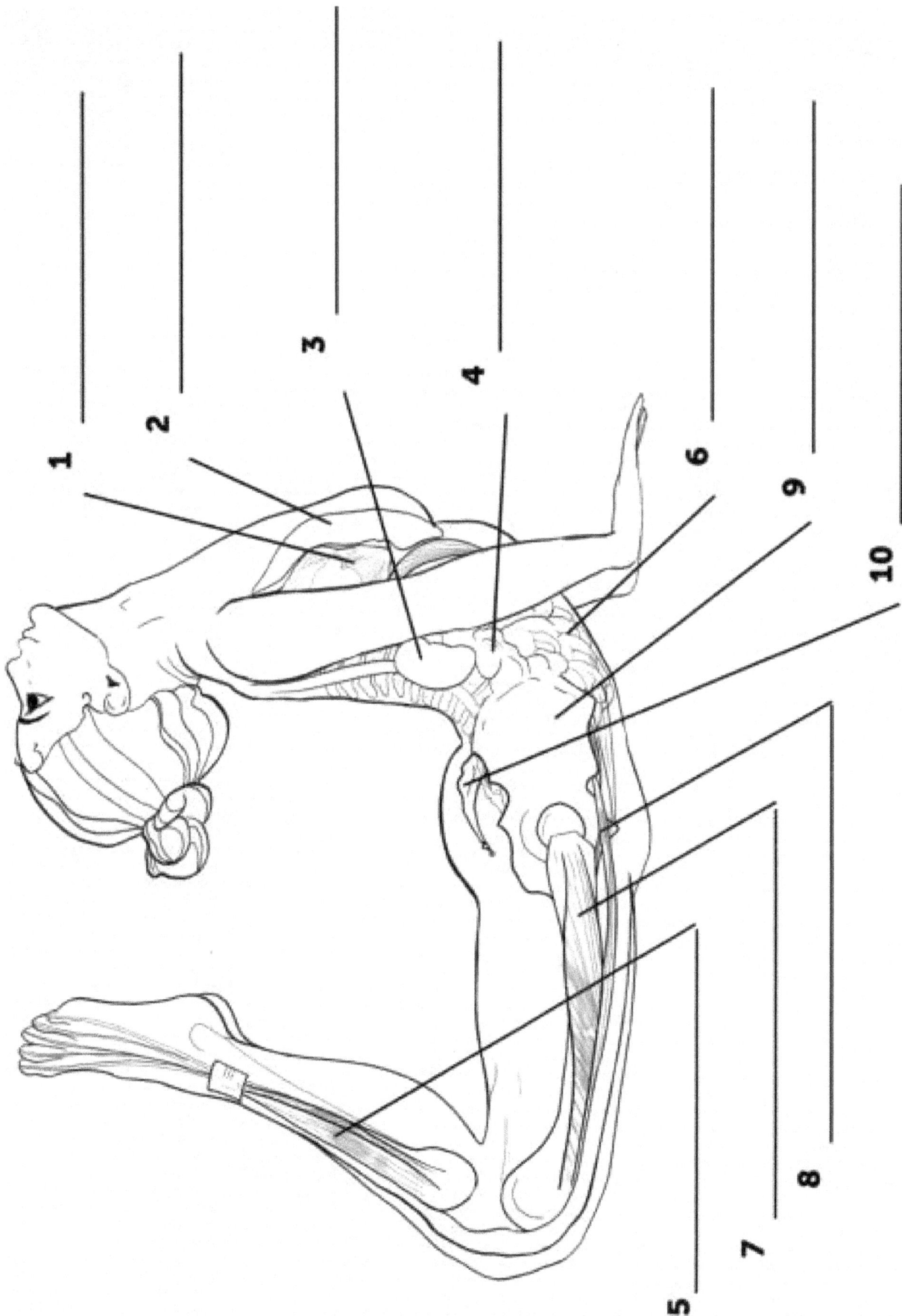

1

2

3

4

5

6

7

8

9

10

155. KING COBRA POSE

1. HEART
2. LUNGS
3. KIDNEY
4. ASCENDING COLON
5. TIBIALIS ANTERIOR
6. COILS OF SMALL INTESTINE
7. RECTUS FEMORIS
8. SARTORIUS
9. PELVIS
10. SACRUM

156. AWKWARD POSE

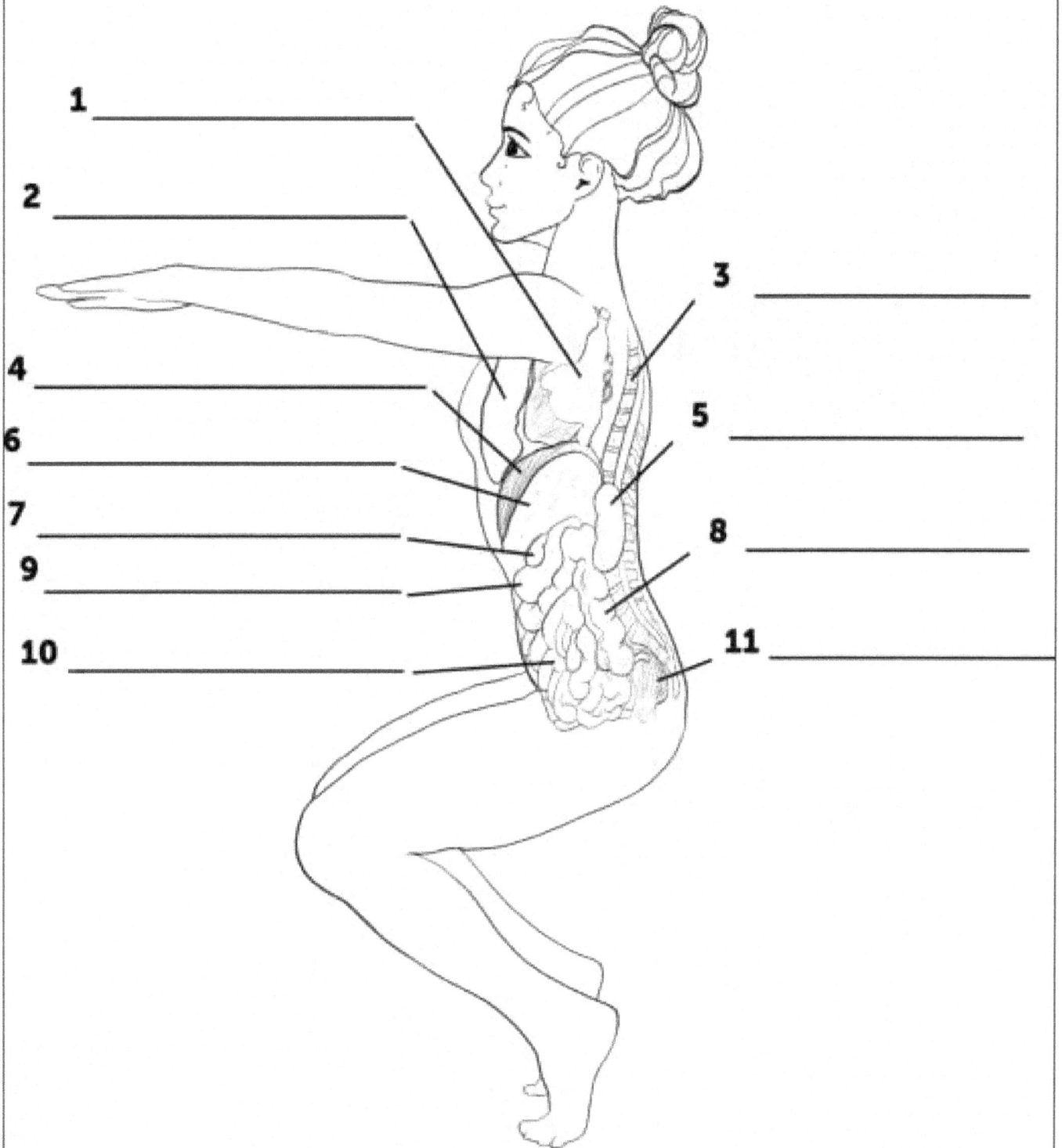

1 _____

2 _____

3 _____

4 _____

5 _____

6 _____

7 _____

8 _____

9 _____

10 _____

11 _____

156. AWKWARD POSE

1. HEART
2. LUNGS
3. SPINE
4. DIAPHRAGM
5. KIDNEY
6. LIVER
7. GALLBLADDER
8. DESCENDING COLON
9. STOMACH
10. COILS OF SMALL INTESTINE
11. RECTUM

157. STANDING HEAD TO KNEE

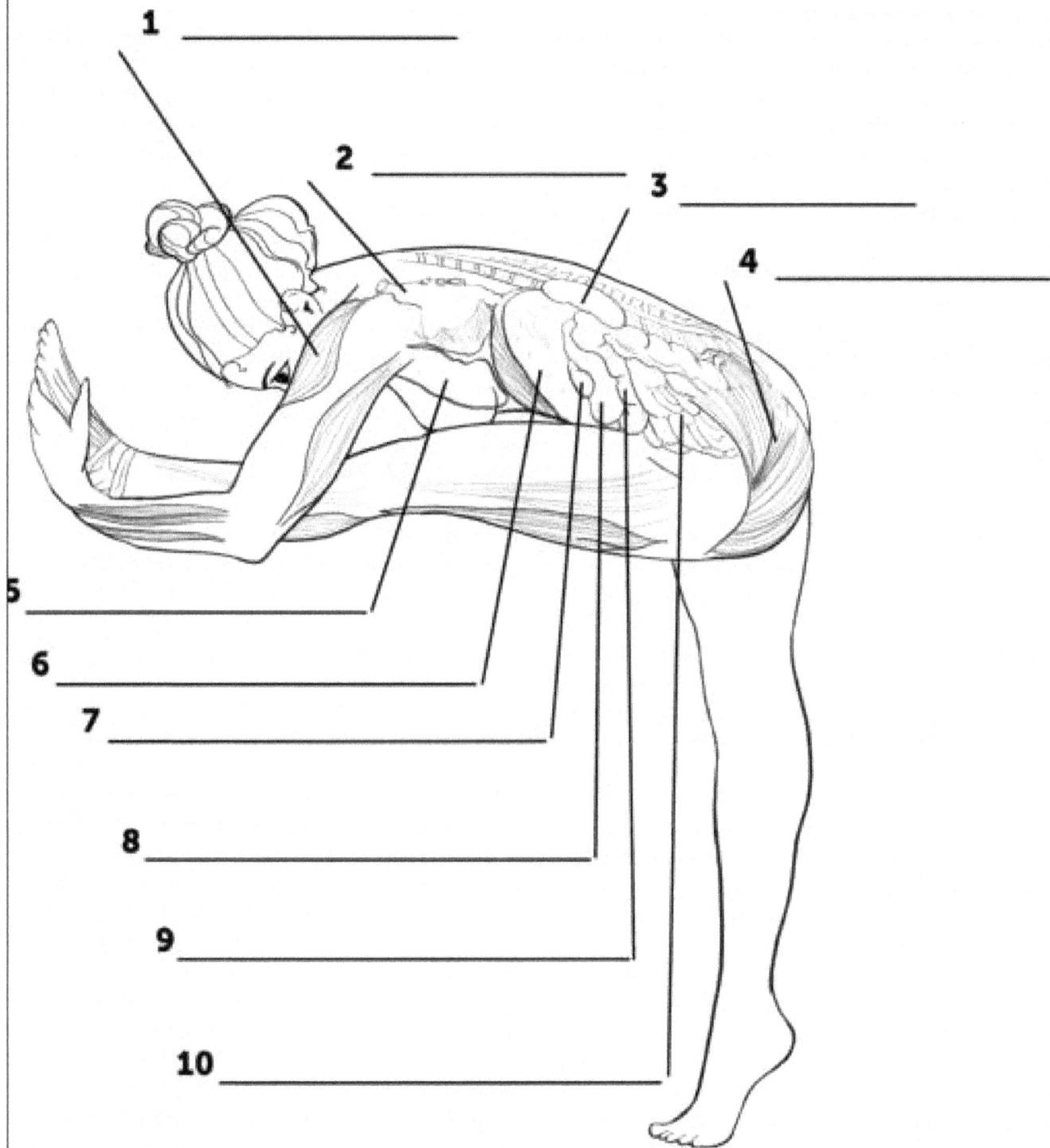

1 _____

2 _____

3 _____

4 _____

5 _____

6 _____

7 _____

8 _____

9 _____

10 _____

157. STANDING HEAD TO KNEE

1.   DELTOID
2.   HEART
3.   KIDNEY
4.   PIRIFORMIS
5.   LUNGS
6.   LIVER
7.   GALLBLADDER
8.   STOMACH
9.   TRANSVERSE COLON
10. COILS OF SMALL INTESTINE

158. UNSUPPORTED SHOULDER STAND

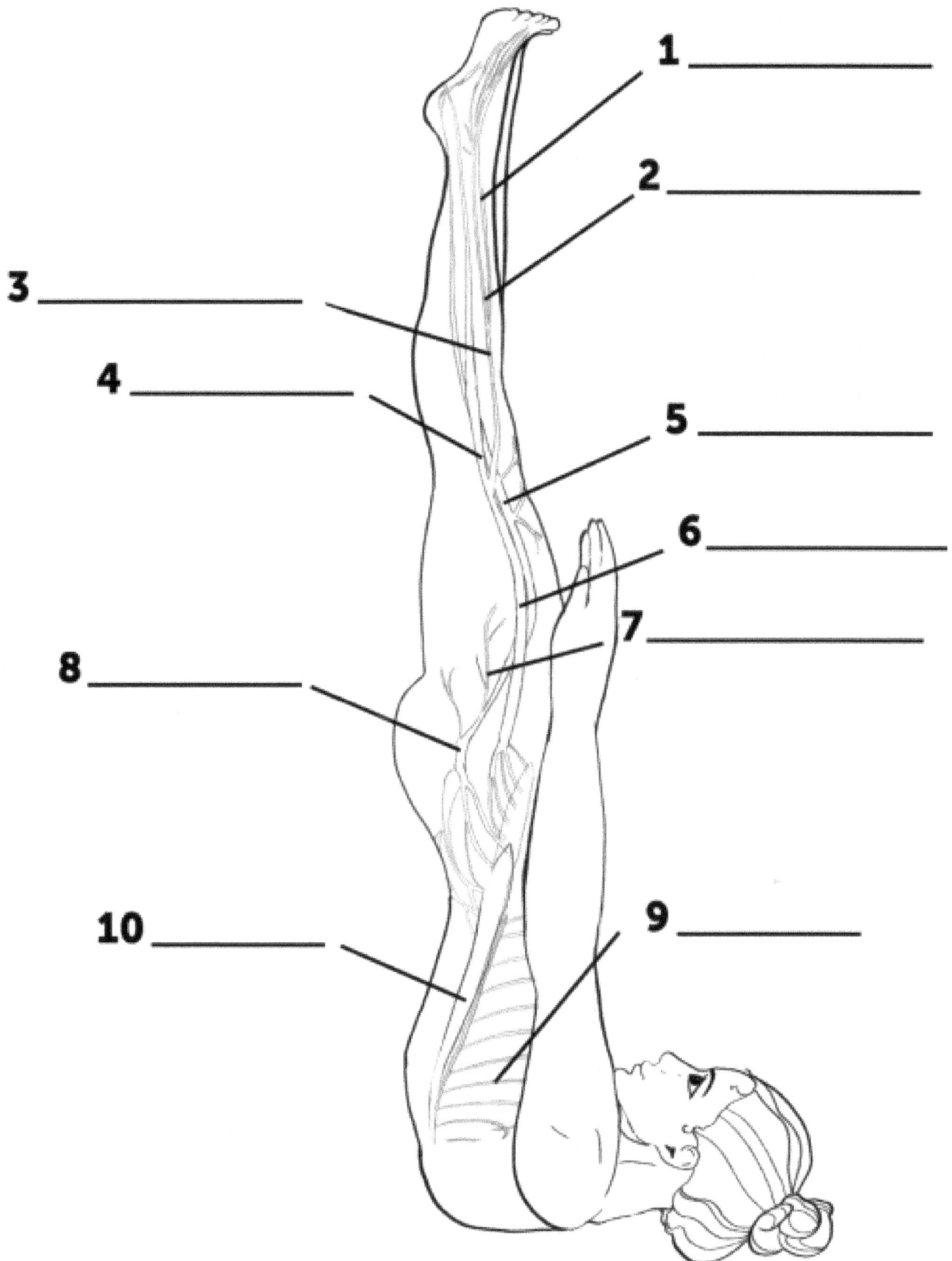

1 _____

2 _____

3 _____

4 _____

5 _____

6 _____

7 _____

8 _____

9 _____

10 _____

158. UNSUPPORTED SHOULDER STAND

1. SUPERFICIAL PERONEAL

2. DEEP PERONEAL

3. COMMON PERONEAL

4. TIBIAL

5. SAPHENOUS

6. SCIATIC

7. MUSCULAR BRANCHES OF FEMORAL

8. FEMORAL

9. INTERCOSTALS

10. SPINAL CORD

159. SKANDASANA

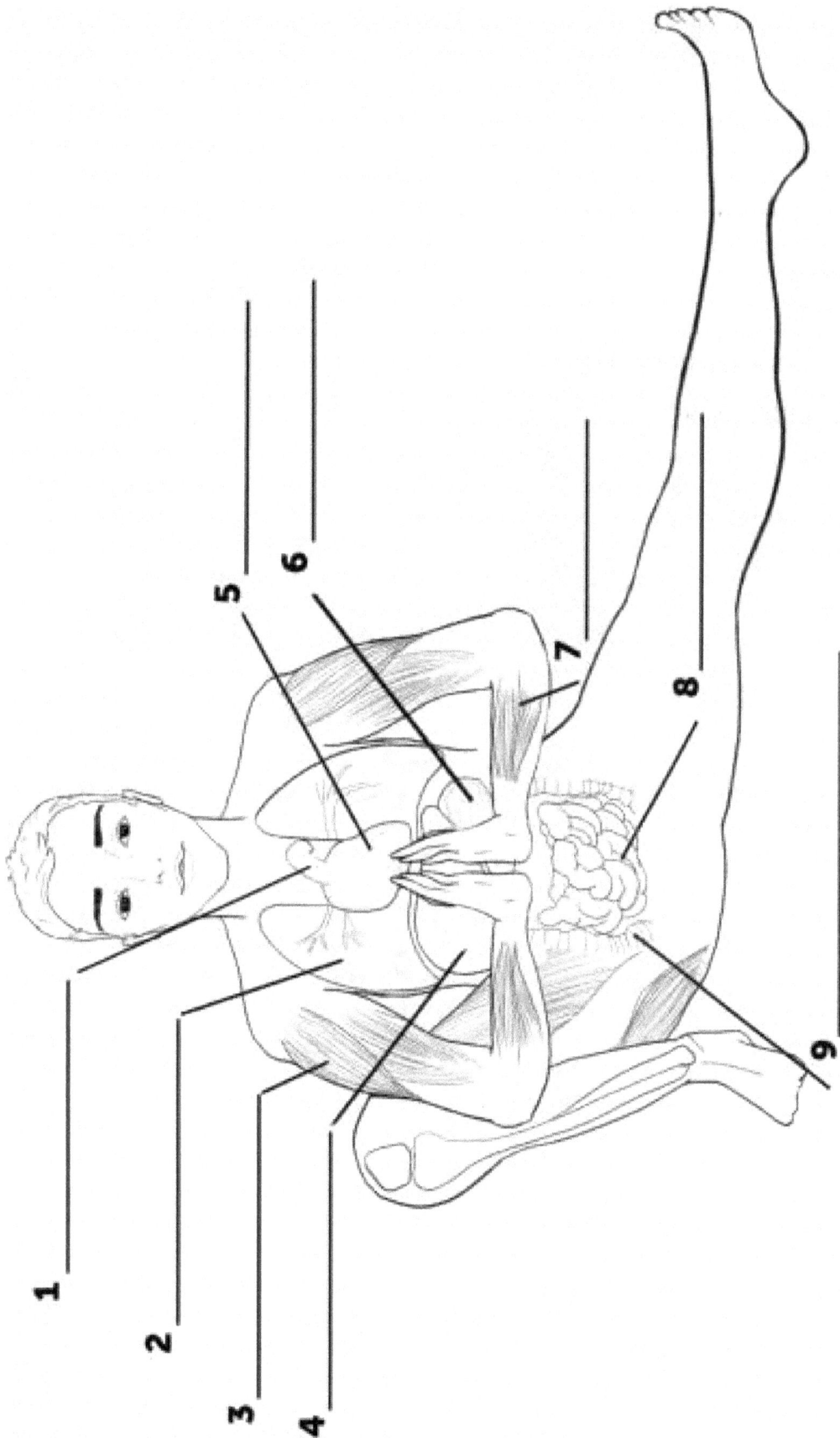

1

2

3

4

5

6

7

8

9

159. SKANDASANA

1. AORTA
2. LUNGS
3. DELTOID
4. LIVER
5. HEART
6. STOMACH
7. PRONATORS
8. COILS OF SMALL INTESTINE
9. ASCENDING COLON

160. SIDE-RECLINING LEG LIFT

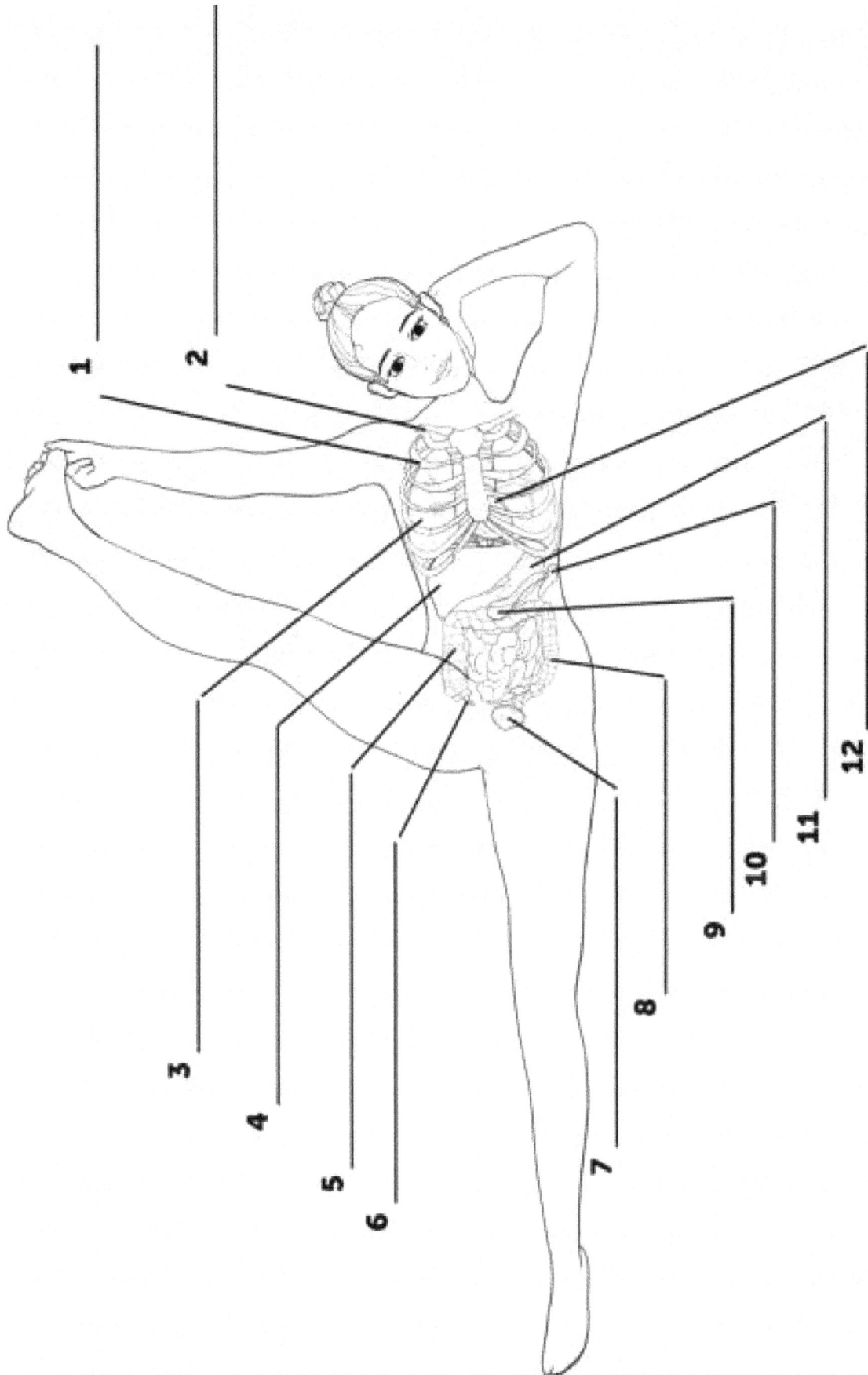

1

2

3

4

5

6

7

8

9

10

11

12

160. SIDE-RECLINING LEG LIFT

1. RIBS

2. COLLARBONE

3. LUNGS

4. LIVER

5. ASCENDING COLON

6. APPENDIX

7. URINARY BLADDER

8. DESCENDING COLON

9. PANCREAS

10. SPLEEN

11. STOMACH

12. HEART

www.ingramcontent.com/pod-product-compliance
Lightning Source LLC
Chambersburg PA
CBHW051333200326
41519CB00026B/7411